PRAISE FOR

The New Bread Basket

"Amy Halloran is right on target. In *The New Bread Basket*, she not only brings all of us up to date on the most exciting new development in the world of grain—local and regionally specific farming, milling, and baking—but also introduces us to a whole new generation of budding, dedicated superstars who are making it happen. This is living tradition at its finest, reinvented in an exciting, contemporary manner. An idea whose time has come—again."

—**PETER REINHART**, author, *Bread Revolution*

"In *The New Bread Basket*, Amy Halloran immerses herself in the burgeoning local grain movement, finding the farmers, wheat breeders, millers, bakers, and brewers who are creating a new food culture. No one has explored this landscape with the depth and passion she brings to the subject. If you're curious about the future of bread, beer, or even the locavore movement itself, this is the place to start."

—**SAMUEL FROMARTZ**, author, *In Search of the Perfect Loaf*

"Halloran profiles a group of thoughtful, committed citizens who are helping to change the world . . . of grains. Beautifully capturing the descriptions of their mannerisms and ways, this book gives us reason to be hopeful that everyday people will heal the planet and our food system."

—**AMBER LAMBKE**, executive director, Maine Grain Alliance

"*The New Bread Basket* is a love story about grain and the people captivated within its embrace. It is about community, connections, and conversations. From the history to the science to the passionate individuals and organizations involved, Amy's book will open your eyes to the revelations taking place every day in the name of grains. Expressive and eloquent, her writing reveals a personal journey that both informs and inspires. As a baker, I give Amy's book my highest recommendation. Reading it is truly a pleasure!"

—**CIRIL HITZ**, master bread baker and author, *Baking Artisan Bread*

"Amy Halloran has created a baby book of the local grain movement and included snapshots of its older siblings: farming, milling, baking, brewing, and oven building. What an amazing family, capable of hard work, persistence, and a generous exchange of information that has helped bring these ancient arts back onto our contemporary food scene. Amy explains how these symbiotic relationships have transformed the flavor, nutrition, resilience, and economies of the local grain movement."

—**RICHARD MISCOVICH**, baking instructor
and author, *From the Wood-Fired Oven*

The

NEW
BREAD
BASKET

The
NEW
BREAD
BASKET

How the New Crop of Grain Growers,
Plant Breeders, Millers, Maltsters,
Bakers, Brewers, and Local Food Activists
Are Redefining Our Daily Loaf

AMY HALLORAN

CHELSEA GREEN PUBLISHING
WHITE RIVER JUNCTION, VERMONT

Project Manager: Bill Bokermann
Project Editor: Benjamin Watson
Copy Editor: Laura Jorstad
Proofreader: Helen Walden
Indexer: Lee Lawton
Designer: Melissa Jacobson

Printed in the United States of America.
First printing July, 2015.
10 9 8 7 6 5 4 3 2 1   15 16 17 18

Chelsea Green Publishing is committed to preserving ancient forests and natural resources. We elected to print this title on 100-percent postconsumer recycled paper, processed chlorine-free. As a result, for this printing, we have saved:

43 Trees (40' tall and 6-8" diameter)
19 Million BTUs of Total Energy
3,680 Pounds of Greenhouse Gases
19,964 Gallons of Wastewater
1,337 Pounds of Solid Waste

Chelsea Green Publishing made this paper choice because we and our printer, Thomson-Shore, Inc., are members of the Green Press Initiative, a nonprofit program dedicated to supporting authors, publishers, and suppliers in their efforts to reduce their use of fiber obtained from endangered forests. For more information, visit: www.greenpressinitiative.org.

Environmental impact estimates were made using the Environmental Defense Paper Calculator. For more information visit: www.papercalculator.org.

Our Commitment to Green Publishing

Chelsea Green sees publishing as a tool for cultural change and ecological stewardship. We strive to align our book manufacturing practices with our editorial mission and to reduce the impact of our business enterprise in the environment. We print our books and catalogs on chlorine-free recycled paper, using vegetable-based inks whenever possible. This book may cost slightly more because it was printed on paper that contains recycled fiber, and we hope you'll agree that it's worth it. Chelsea Green is a member of the Green Press Initiative (www.greenpressinitiative.org), a nonprofit coalition of publishers, manufacturers, and authors working to protect the world's endangered forests and conserve natural resources. *The New Bread Basket* was printed on paper supplied by Thomson-Shore that contains 100% postconsumer recycled fiber.

Library of Congress Cataloging-in-Publication Data
Halloran, Amy, author.
  The new bread basket : how the new crop of grain growers, plant breeders, millers, maltsters, bakers, brewers, and local food activists are redefining our daily loaf / Amy Halloran.
    pages cm
  Includes index.
  ISBN 978-1-60358-567-5 (pbk.) — ISBN 978-1-60358-568-2 (ebook)
1. Bread. 2. Bread—History. 3. Grain. 4. Grain—History. I. Title. II. Title: How the new crop of grain growers, plant breeders, millers, maltsters, bakers, brewers, and local food activists are redefining our daily loaf.

 TX769.H233 2015
 641.81'5—dc23

                    2015009574

Chelsea Green Publishing
85 North Main Street, Suite 120
White River Junction, VT 05001
(802) 295-6300
www.chelseagreen.com

To the long line of people and plants
who collaborate to feed us, and
to one collaborator in particular,
my son Francis.
His curiosity about the
intersections of humans and the land
is one place this book began.

CONTENTS

INTRODUCTION

The story of wheat is the story of everything. How we get our staple crops defines who we are. Potatoes, wheat, and rice: The things that feed us can show our connection to land and machines and to one another, or lack thereof. A bag of potato chips offers a lot of crunch, but that noise is mum about the Great Famine. A million Irish died and more than a million emigrated because the island nation planted only two varieties of the potato. Abandoning the genetic diversity found in South and Central America, where the crop originated, spelled disaster. Despite the link to starvation, colcannon and other potato-based dishes still define much of Irish cuisine.

How we feed ourselves feeds our imagination. If I were a country, pancakes would be my national dish. I fell for flour as soon as my mother began to teach me fractions on measuring cups. The power of flour was in its alchemical nature. Take this magic powder, mix it with water and other simple ingredients, bake it, and it becomes delicious. As an adult I endeared myself to people by baking them birthday cakes because I wanted them to taste my affection. I bought good ingredients, but I didn't think much about flour until I was in my forties, when I tasted a certain cookie made with oats and wheat that had been grown, rolled, and milled near where I live in New York State.

My husband brought me an oatmeal ganache bar on his way home from a business trip. I was skeptical about the gift. Little did I know how many worlds that cookie would open. Even against a backdrop of good butter and chocolate, I could really taste the grains. Their flavor and freshness introduced me to the regional grain revival that was happening right under my nose, and sent me on an investigation, one that has culminated in the book you're now reading.

Nowadays I spend much of my time considering wheat and the history we share with the edible seeds of certain grasses. I think about how planting wheat and other foundational crops helped settle our nomadic ancestors. I think about the processes involved in growing grains and turning them into food. I wonder why those processes have become so invisible.

There is a long line of people, living and dead, standing between me and my favorite ingredient. The particular history of each grain is a microcosm of the general history of farming, milling, and baking. I love to peer into these windows, reading about people storing grains in reed baskets and mud pits. Much more recently, and closer to me, people ferried grains across the Hudson River to be ground into flour by a water-powered mill. Along the banks of the Poestenkill, millstones sit among the other rocks as evidence of that era.

Ghosts of wheat sit between these forgotten millstones and the bags of flour on a supermarket shelf. If we could see all these ghosts, understand the steps that led farming, milling, and baking out of sight and out of mind, would we be as dubious of wheat and gluten today?

Wheat is my favorite storyteller. More of the world is planted in wheat than any other crop. At times, wheat has been the primary fuel for everyday life. Between 1900 and 1940, bread made up 30 percent of our calories in the United States. Grains and bread are central to our eating and experiences. We don't break butter or apples, we break bread. *Our daily bread* is shorthand for simple sustenance and spiritual nurturing, yet most of us can scarcely see the amber waves of grain so lauded in our national hymn, "Oh, Beautiful."

As people work to regionalize food production, staples like grains are the last piece of the locavore puzzle to be solved. The relative stability of grains, which since ancient times has made them a good food to store, is the same thing that has allowed this staple to become a commodity, vanishing into the anonymous middle of the country.

Whole villages used to drop everything and gather together for the grain harvest. Yet the mechanization of harvesting equipment and baking systems gradually broke the cohesive property of this crop. Once you know how grain production and handling have changed, a bag of sliced white bread can speak volumes to you. Long an easy shorthand for social and dietary woes, soft bread, and its tanner whole- and multigrain cousins, might begin to explain wheat sensitivities and rising rates of celiac disease.

Some say bread helped civilize us, because we had to sit still to grow grain, and grain freed us to pursue things that were not essential to the business of being alive. Others argue that agriculture is the great mistake that created social stratification and began the environmental degradation that threatens our personal and planetary health. While we may be headed for an apocalyptic future, I'm betting that bread and beer can tame the human animal again, and help us come to peace with the staff of life, as well as with the even greater things we rely on, like air and dirt and rain.

In the four years since I ate that door-opening cookie, I've seen a lot of people working to rebuild regional grain systems. In the process they are reviving economies, relationships, and communities. Decentralizing production of staple crops requires cooperation, and grain projects are rebuilding more than markets and infrastructure. The people are recapturing the social meanings of wheat and restoring the real value of grains to our lives and the land. This book is my salute to the people who are making and breaking bread, and brewing a village into every pint. I hope you will enjoy this tour of their passions.

CHAPTER ONE

THE FARMER, THE MILLER, AND THE BAKER

"We started this flour mill before people got excited about local grains," said farmer Thor Oechsner, looking down at the line of cars stopped in front of his red grain truck. "Now people come to look at us as a model. They tell me we're at the forefront of the grain movement, but it doesn't feel like that from here."

Growing wheat outside the grain belts might seem revolutionary, but the work itself is like most other jobs, full of interruptions and crowded with tasks. Thor was picking up wheat from an organic dairy farmer and bringing it to the mill. As construction crews repaired a bridge, traffic was reduced to a single lane. The October sky was wide open, and a ring of perfectly exhaled clouds offset the pale oranges and bleached yellows of the trees. Fields were green with alfalfa or tan with drying corn.

My love for flour landed me in the cab of this truck, and in the middle of the regional grain revival. While millions of Americans today shun gluten, others are rebuilding their lives around bread, like this farmer and the millers and bakers he works with outside Ithaca, New York.

A chatty, generous fellow, Thor invited me to visit after we met at a conference. He's shown me wheat sprouts fingery green in the fall, and stalky tall and golden, ready for summer harvest. I've sat in the cab of his combine and watched the machine munch that harvest like a greedy bug, skittering over the field and filling its great metal belly with grain, spitting straw out the back.

I also have front-row seats at a bakery and mill because Thor is part-owner of Farmer Ground Flour and Wide Awake Bakery. I've marveled at the Rube-Goldberg-esque mesh of tubes and machines that Greg Mol, Neal Johnston, and Benji Knorr use to grind flour. Bakers Stefan Senders, Hope Rainbow, and David McInnis have even let me wear an apron and try to help.

All the while, I looked at the wheat in the field and the kernels in the bins, and I thought, *Are you going to feed me?*

Amber waves of grain are words in a song. They don't even make pictures in my mind. When I hear them in my head, sung by an anonymous but perfectly tuned chorus, I think of the flag and vague ideals of patriotism. The phrase is more evocative of America's hopes for itself than the stuff that becomes the bread I butter. Yet the story of wheat is the story of everything.

Our staple crops matter deeply. They are the roots of our diets, central to our health and ability to flourish. Potatoes once swelled the population of Ireland by five million people in less than a century. When blight hit and killed the potatoes, one million people who depended on them also died, and more than a million left the country. Yet the tuber today remains a favored Irish food. My father, whose great-grandfather left the island because of the famine, could eat mashed potatoes at every meal.

How we eat is ingrained in us as individuals, and in our cultures. For millennia, bread provided the bulk of our calories. That's why *breaking bread* means more than just sharing food. When we work for a living we are *bread-winners*, and *dough* often means "money." These terms are so ubiquitous that they have become almost invisible metaphors. And the process of growing bread is almost invisible, too. Pictures of wheat stalks grace cracker boxes, but most people don't know the first thing about grains or how we grow them.

As I've followed flour back to the field, I've loved and questioned everything I've seen: "That's the kind of wheat that makes my favorite flour?" "That's the way a combine jitterbugs?" I hate heights, but I have climbed the ladder on a grain bin out at Thor's farm and looked down at the veritable sea of stored grain.

Here is what I see. In the midst of fields and rows of trees and other farms, Thor's farm fits a standard image, with barns and tractors and a small farmhouse. There are rows of grain bins, their corrugated steel skins shiny as tinfoil, their tops cone-shaped like the Tin Man's hat in *The Wizard of Oz*. Another big-bellied tank sits under a roof, ready to dry grains for storage. Three newish white buildings, tall with green roofs, hold equipment and a workshop. In the shop, Thor works on the small fleet of tractors he keeps. There are nine in all, and two combines. Next to the shop is a seed cleaning room. The other buildings hold mowers and balers and plows, the tractor attachments that ready fields for cultivation, planting, and harvest. Behind and around the machines are lots of totes of grain, white cubes sewn of plastic fabric, tall as my shoulder, and sitting on pallets, ready to go to malthouses, distilleries, and mills. Another barn, a classic with gray, weather-beaten siding and a hayloft, holds hay. This is the oldest building on the property, probably dating to 1860.

Over the years I've known Thor and the farm, the storage bins have grown like mushrooms. Another row popped up by the seed room, drawing a line between the grain buildings and the paddock and barn where Rachel, Thor's girlfriend, keeps her horses. Thor and Rachel have a flock of uncountable chickens, and I love these eggs more than any others in the world. The yolks are the brightest orange, because the birds roam around the farm, feasting on grubs, kitchen scraps, and treasure-trove spills of organic grain.

I never tire of visiting. I love learning which grains grow well where. What makes a good growing season. What makes a bad harvest. The details make a map, and in my mind I can walk a seed from ground to griddle as I make pancakes—my favorite food.

Sometimes a pancake is just breakfast, but many days I'm in awe. That any little seed can fill our bellies is stunning enough, but the people and processes that add up to flour and bread are the real stars. The six-thousand-year-old habit of leavened bread fosters all sorts of communion. Break down my pancake, or a loaf of bread from Wide Awake Bakery, and you'll see the human glue that binds together farmers, millers, and bakers.

Real bread takes time. As I sat in the red grain truck that fall day, waiting for a construction worker to flag us forward, the next year's bread was already growing. Wheat and rye are often fall-planted, and by October the fine fingerling hairs of seedlings make fields look like they are covered in green blankets. The hairs thicken into little clumps and go dormant in winter, ready to shoot up high once the weather warms in the spring. The sourdough starter that builds bread was under construction, too, sitting in a ceramic pot at the bakery. In other words, the toast I ate on an October morning was planted in September the year before.

When I first figured that out, I was floored. I've been a baker forever, but I didn't start thinking about flour until a few years ago. Sure, I knew to buy quality flour, and used King Arthur exclusively, but I didn't think beyond the brand. I certainly didn't think about the length of time between seed and loaf.

Every step is all new to me, even as it grows familiar. The languages of grain farming, milling, and bread baking each have their own elaborate vocabularies. I will always forget how to spell the word *vomitoxin*, a poison caused by *Fusarium* fungi, and I can't remember if ergot comes from the same fungus. (It does not. Another fungus, *Claviceps purpurea*, causes ergot, which can create hallucinations in those who eat it, and is blamed for bewitching young women in colonial Salem, Massachusetts.) Just when I think I understand the history of milling, I learn something else and am reminded again of the complexity of plants and machines. What I know is always crippled by the limits of my surroundings. No matter how enchanted I am with food production, I am an eater and a writer, and spend most of my time staring at a computer.

One particular day showed me how grossly uninformed I am. Thor took my kids and me to a farm auction. The tools and equipment that ran a dairy farm for eighty years were laid out in a field. Hay wagons. Grain boxes. Tractors and plows. These are things dairy farmers need to feed their herds.

Thor wanted a set of Norwegian Kverneland plows, the Mercedes of its kind. Inspecting the rows of steel blades, he gave Felix, my youngest, a lesson in naming the parts: frog, share, shin. Landside, trashboard, and moldboard. Felix absorbed and repeated the absurd-sounding names. A set of John Deere plows was up for auction next to it, and Thor compared the two machines. Using his hands, he described how the curves of the coveted plow would scoop up the soil and turn it gently. The steel itself, he said, was of better quality.

As we snaked around the rows of wheelbarrows, chains, and grain dryers, Thor explained the functions of some of the machines. I'd grown up a teacher's daughter in dairy country, but the farms that surrounded us were as good as invisible even then.

At the farm auction, I stuck out like a sore thumb, a woman in a skirt with a notebook. Other women stood under a tent, serving pulled pork and baked beans from Crock-Pots. About sixty farmers followed a pickup truck through the field. The auctioneer called from the cab. Amish or Mennonite farm boys, dressed in long pants and boots on a blazing-hot day, stared enviously at my sons' shorts and sandals. When the truck stopped at the coveted plows, Thor and a couple of other farmers bid with subtle moves that could have been missed: the slightest nod, the tip of a pointer finger to the brim of a cap. I chewed off most of my nails in a few seconds, caught up in the tension of the bidding. Thor paid almost $10,000, nearly six times as much as the John Deere plow that got auctioned off before the one he bought.

The amount of money that used farm machinery cost surprised me, but what lingers about the event is how removed agriculture is from everyday life. The auction fanned out pieces of a dead farm to other farms in an efficient distribution of parts, almost like an organ transplant. We need a bigger surgery, though, for America and farms. We have amputated the social limb of farming from the body of our country, relegating agriculture to an unimportant rank, tucking the work out of sight and largely out of mind. Re-localizing food production is helping change that, but very, very slowly. The more I see of fields and equipment, the more farmers and food processors let me in on their vernaculars, the more I know that we, the eaters, don't understand. How will we close the gap between our mouths and the people who feed us? We watch Olympians train. Study candy factories on TV, eager to see how sugar spins into Snickers bars. How come we don't know the first thing about the flour that makes our bread?

Whole villages used to leave their tasks for the grain harvest. Now we have more people living in prison than growing food on the land. I do not know what level of technology is right to feed our population, but my own family's experiments in urban homesteading left me exhausted. Preserving the food we grew ourselves and what we gleaned from farms took most of our time, yet we were nowhere near self-sufficient. Compared with combines and other tools of small-scale grain farming, snipping wheat stalks with garden shears seemed ridiculous. As I eat and wonder about how involved everyone should be in food production, this farmer-miller-baker collaboration is a good example to study. I am grateful that these people who are so engaged in their work let me stand nearby, asking questions.

"Thor was right there at the very beginning, sitting around the dinner table eating bread," Stefan Senders remembered. "He said, 'This rocks, let's make it happen.'"

Stefan, an anthropologist and home baker, was enticed by the idea of making a bakery, but doubted it would pay. Still, his life was taking a turn, and he was passionate enough about baking to explore the reality of a fantasy he'd entertained for years. He sat up all night crunching numbers, and by morning found that running a bakery could possibly work.

"I went to Thor and said, 'Let's do it,' and he wrote me a check right there," Stefan said. Thor was not alone in supporting the vision, and I can see why. Stefan is a persuasive person. If he has an idea, he'll wrap you inside it, hugging your brain and your heart.

People bought shares in the bakery concept, financing the construction of the building and its huge brick oven. Some people invested without expectation of return. Others bought bread futures, purchasing 250 loaves for $1,000. Others, like David McInnis, bought into the project body and soul, carrying bags of mortar and sweeping the floor while Stefan built the oven with Billy O'Brien, who wasn't a mason but a mechanical genius and self-taught engineer.

Solid as the oven is, it represents something I can't see, the support of many individuals who decided this bakery belonged here. This bakery began as a thought that Stefan embraced, a thought that he and others hoisted into place. The oven also represents another invisible mesh: the conversations Stefan had had with people around the world as he consulted and quizzed them, trying to figure out the best way to make the best oven. Strangers Skyped with him, and spent hours on the phone. Stefan and Billy made an oven seemingly from thin air, harnessing Stefan's ideas of another life, made of bread. The men built a brick firebox under the rotating deck, a little room to trap flames and build heat. On the other side of the structure, a few feet off the ground, the oven's mouth is a metal door that slaps open like a tongue. Bakers feed loaves into the heat using a long wooden peel.

The bakery is woven to whole communities, to all the people who talked to Stefan about bread and ovens, and to the locals who believed in this place when it was just an idea. The bakery is connected to all the effort that's gone into every bakery in history, and every loaf that's ever been made. The generosity and information shared by bakers and oven builders, and the belief expressed by community funding—these intangibles are embedded in the bricks and mortar. Some argue that bread built community, settling hunters and gatherers to grow grains and, eventually, grow civilization. There's no doubt that community built this bread.

Wide Awake Bakery sits in the middle of farmland. The oven sits in the middle of the bakery and takes up most of the room. Bakers standing at hip-high wooden tables, shaping dough into loaves and staring out the windows, might see sheep or buckwheat or horses.

Most of the breads are sourdoughs, so the process of making bread actually begins two days before baking. Fermentation starts on day one. The doughs are then mixed, fermented, and shaped on day two, and ferment slowly overnight. Sometimes the room is quiet, as doughs rise in the cooler and the oven warms, the hum of a bake about to pounce. Later, the space wakes up. Batches of dough are divided and shaped, tucked in baskets and stacked on trays on a rolling cart, ready to be fed to the oven. Bakers flip baskets onto peels, wooden paddles with long handles. They score each loaf with a razor blade and somehow shuffle the unbaked loaves onto the oven's deck. Cranking a handle near the door, bakers rotate the deck and fill the oven.

Soon the dough is bread, unloaded onto racks to cool. People stop by for fresh loaves, and bakers load boxes and baskets for distribution. Cars take the bread away and the bakery quiets again, a shell ready to turn flour into food. Everyone rests until the bakers and bakery once again waltz the grain from the fields through a dance, and make bread.

The fuel that feeds this bakery, a community-built enterprise, is still people. This may be said of any business, but the relationship here is more direct. Bread is more than a shorthand term; it is a currency in this system. Wide Awake Bakery runs as a CSB, a community-supported bakery. People contribute to the Crust Fund, buying a ten-week Breadshare for $50. Instead of carefully grown produce, people buy subscriptions for carefully made loaves. Crust Fund numbers hold steady throughout the year at about 250, but more people eat the bread in the summer because of auxiliary subscriptions, add-ons to area CSA farm shares.

Once a week, a bake of bread and sweets and pasta goes to the farmers market in Trumansburg. These go to a few stores, too, and to CSB pickups. People covet this bread. They greet it and the bakers with gusto. Some buy extra loaves—one to devour on the way home, and more to share at a more moderate pace with their family.

"A lot of people are proud of this bread," said bakery manager Hope Rainbow, sitting on the grass at the farmers market. This was her day off, but she came to the market anyway. When she requested an apprenticeship a few years ago, she wrote to Stefan that she knew the bread because friends felt compelled to share it. "Try this!" they told her. "This is our bread, made with local flour, made from our local wheat."

That pride was appealing. The bread was riveting, and so much less abstract than her studies as an art history major. When she graduated, she wanted to be a part of something substantial. Now she is.

That substance is symbol, too. The bakery has sent loaves to anti-frack-ing rallies, beautifully echoing a Bulgarian bread march that, along with other protests, led to Bulgaria becoming the first country in the world to ban shale gas extraction. New York State was not so easily convinced to shut out the fossil fuel companies. "Break Bread, Not Shale!" people chanted at the state capital and elsewhere. For two years, environmental activist and poet Sandra Steingraber held loaves in the air as she eloquently listed reasons why hydrofracking shale to extract natural gas was dangerous. Finally in late 2014 Governor Andrew Cuomo banned fracking in the state, but environmental threats to the Finger Lakes region, and the viability of its farm economy, continue. Steingraber and her baker Stefan were among many arrested for blockading the construction of underground gas storage at Seneca Lake.

Here's a snapshot of how the farm fits in the bakery, and how the mill and grain farming fit into the bigger picture of grains in America. About 70 percent of the flour used in the breads at Wide Awake Bakery comes from Farmer Ground Flour. This mill processes 30 tons of grain a month, and some of that is custom milling. Figuring a ton or a ton and a half per acre, that's close to 400 acres each year.

Bread flour is the mill's most popular product. Working organically, Thor has a rotation system to avoid buildup of diseases and pests. Though he farms up to 1,200 acres, in any given year only a few hundred might be in wheat. Crop rotations are a standard practice for organic farmers, building soil health as a defense against the kinds of crop threats that conventional farmers fight with inputs like pesticides. Thor likes to make long rotations, cycling through grasses and legumes over the course of six or seven years, and limiting how many acres he can plant to each grain. Drainage and other soil conditions also limit how much ground can be planted to wheat. The acreage he plants is spread out over a 15-mile radius, complicating the shuffle of machines and work.

The climate in the Northeast is not perfect for grain farming. Humid summers mean lower protein levels, which means that grain crops can miss the performance qualities bakers want. Moisture in the air can spell trouble as the kernels mature. The risk of contamination with fusarium head blight, a fungus that causes vomitoxin, is higher here than in drier areas. Vomitoxin levels must be below 1 part per million for food-grade grains. Grains that fall short of food grade might still be good for animal feed, or for making alcohol, but not for bread.

Elsewhere, bread wheat grows in more concentrated areas. Many wheat farms are 2,000 to 5,000 acres in size, and fields can be flat and huge, not chopped up by hills and houses. The wheat belts of the United States tend to have dry summers, which boosts protein in wheat. Bakers have grown to expect high proteins in bread wheat, which creates a challenge for farms and mills in the Northeast.

Given these constraints, a diversity of plant species is necessary for a healthy crop rotation, which necessitates a diversity of markets. Thor sells hay and feed-grade grains to organic dairies and chicken and hog farms; he also sells rye, barley, wheat, corn, and buckwheat to maltsters, brewers, and distillers. He also grows buckwheat, oats, and red clover for seed markets. Having many customers is essential to the financial success of Oechsner Farms, and the mill needs more than his grain to meet demand. So Thor coordinates with other farmers, contracting with them to grow soft wheat for pastry flour and hard wheat for bread flour. If his rye crop doesn't yield what he anticipated, he has to get on the phone and find people who have good, food-grade rye to sell to the mill. Greg Mol will be taking over this work of brokering grains as systems fall into place at the expanding mill, but for now the job is Thor's. He says he's too busy, and he probably is. Yet he is also a compulsive worker who thinks of retirement as farming only 500 acres. This is like saying you'll work a few days a week, instead of most of them.

Thor was born in the suburbs but caught the farming bug from visits to his uncle's farm in Pennsylvania, and a friend's farm in Ohio. As a teenager, he convinced his parents to let him tear up the lawn and plant corn. This was in Westchester County, just a commuter train ride north of New York City. His corn stood out in the stately residential area, just as his enthusiasm for grain farming still leaps out at anyone who meets him.

The idea for the mill hit him as he watched the land he rented get swallowed up for housing. He knew he needed to make more money than he could by growing organic feed grains for dairies. He thought of the farm where he had worked as a teenager, a farm that survived in a high-priced area by opening a slaughterhouse to help market its livestock. Thor knew he had to add some value to his crops. Milling was the first thing that came to mind, because his grandfather was a baker.

Like the bakery, the mill began in conversations. Thor heard that another farmer, Erick Smith, was also interested in milling. The two decided to work together instead of competing. Thor bought a 24-inch stone mill he found through an ad in the back of a farming newspaper. The mill sat in his barn for a year and a half because the two farmers were too busy to set it up. Besides, the enterprise was a puzzle. A hundred years ago, there were plenty of mills around. Nearby there was an artifact at the Robert H. Treman State Park. The water-powered mill was nice to look at but wouldn't say much about running a small mill in the twenty-first century.

Vermont grain farmer Jack Lazor was helpful. Erick and Thor visited Butterworks Farm, the dairy Jack runs with his family. They saw how he used a small Meadows mill in a barn, and heard how he sold his flour, mostly to food co-ops. (Meadows is a company in North Carolina that manufactures stone mills; it is the main outfitter of mills in America for small-scale use.) Jack was encouraging, but his setup wasn't a blueprint. The prime product of Butterworks Farm was yogurt. Erick and Thor grew staple crops. Figuring out how to add a mill to their operations remained a puzzle, and a back-burnered project.

Luckily, an outside influence arrived. June Russell is the farm inspector for Greenmarket, which runs farmers markets in New York City. As June inspects farms, she keeps her eyes and ears open for producers that can fill gaps in what's for sale at the markets. At the time, no one was selling dry beans, so when she heard that Erick Smith had beans to sell, she canceled a planned stop and went to see him. They discussed what it would take for him and his company, Cayuga Pure Organics, to sell beans in New York City. As June was about to leave, Erick mentioned that he and his buddy had a mill and were thinking of making flour.

"You need to do that immediately," June told him. It was 2008. The word *locavore* had just made it into the dictionary. Stands at Union Square and other markets that Greenmarket ran were bursting with lettuce, kale, and chard. Locally raised meat and eggs went for good money. Yet the flour that bakers used came from far away and had nothing to do with the organization's mission to support farms around the New York metropolitan region.

June was excited. She was hunting for flour to try to get bakers aligned with the mission. But her sense of urgency couldn't make the mill happen. Thor and Erick were busy chasing sunshine and fixing machinery. The idea of the mill was something that just couldn't get done. They wanted to emulate Jack Lazor's simple barn operation, but New York State required a commercial kitchen license, and building a new facility would be prohibitively expensive. Profits in grains and flour are too minimal to support big investments in infrastructure. The two farmers wondered if the mill was really worth pursuing.

"We have bakers who will use your flour. We can guarantee you a market and we will support you," June said as Erick called the office, looking for help and funding. "The time to do this is now."

The nonprofit had a huge market to offer, but not money. From downstate, how much coaching could they provide? The extra oomph came from a local food champion, Gary Redmond.

Gary had been building farms and food businesses in the Finger Lakes since the mid-1980s, linking farmers and food producers to markets by covering the logistics of distribution. Aptly titled Regional Access, the business started in his garage, quickly outgrowing the space and moving to an old Agway feed building in Trumansburg. Eventually, the distributor needed new digs with refrigeration and cooling capacity, along with room for pallets and forklifts.

Gary had a passion for pushing any kind of local food enterprise and was eager to fill the milling gap in the local food system. At one point, he had been involved in Community Mill and Bean, a cooperative that handled staple foods. He wanted to see local milling happen again, and the old Agway building would be a good starter home for the mill. The space was right because it had a history in grains; in addition, the building housed a catering company, which had a commercial kitchen license. The mill could piggyback onto existing facilities, such as a bathroom, rather than outfitting another place from scratch. This made Farmer Ground Flour's licensing process simple at the state level, and affordable for the farmers.

Gary Redmond helped with more than a roof. He didn't charge rent while Greg Mol, a friend of Thor's, built the mill. Gary gave Greg free rein to make noise and dust, even though he lived in an apartment at the front end of the building.

Greg had come to Ithaca to work with organic grain farmers after five years working on CSA vegetable farms in New York and Pennsylvania. His interest had shifted because grains seemed a more substantial issue for sustainable agriculture, and one that was getting less attention. Greg was working for Potenza Organics, the primary soybean supplier for Ithaca Soy, when Thor and Erick approached him about the mill. While he wanted to farm, the mill was too good an idea to pass up.

"If you look at independent mills, there's always a connection to agriculture," Greg said. Family farms turn to milling for control of their crop and income. The idea of Farmer Ground Flour, a mill owned and run by organic grain farmers, really captivated him. Just as farmers markets acted as bridges to build local agriculture, mills are essential infrastructure for leveraging production of staple crops in small acreages and out of the commodity system.

Thor and Erick invested money in the business, and Greg invested sweat equity. He worked part-time on Thor's farm as he started building Farmer Ground Flour. He took a grain milling course long-distance from Kansas State University, and sought help where he could as he laced tubes through the beams and floor of the building. When the mill was ready, he brought 50-pound sacks of grain from the farm in his pickup truck. Hoisting a bag on his shoulder, Greg climbed a ladder to feed the mill's hopper.

Today grain travels from bins at the farm to bins at the new mill, which was built by an Amish family in the fall of 2012. Like the bakery, the mill sits in a field. Tan and boxy, part of the structure is three stories tall. Sections of the second floor are not attached to the outside walls, to ease vibrations from the sifters. Greg and fellow miller Neal Johnston used what they learned at the first setup to streamline operations. Neal customized the electronics for the stone mills and the sifters. The two of them installed seed cleaners. A wall of shelving is stacked with totes of grain ready to mill, and white paper bags of flour. Thor plants the fields that surround the mill to rye, buckwheat, or other crops that might make their way indoors.

The structure of the business changed when they moved into the new building. Erick Smith left the cooperative, and Greg and Thor offered Neal the chance to become a member. Now the three are equal shareholders. (Erick died in the fall of 2014.)

Outside the building, four 1,000-bushel bins hold clean wheat and rye ready to be milled into flour. A fifth bin, known as the dirty bin, holds incoming grain that needs to be cleaned before coming into the system. One day there might be enough outside storage for everything, but for now things like buckwheat and corn that aren't used in huge quantities sit in 1-ton tote bags, labeled with strips of duct tape to tag their contents. The first day that wheat traveled from the bins at Thor's farm to the bins at the mill without having to be bagged or toted up, I was visiting. This was a big leap from Farmer Ground's beginnings. Going from Greg hauling 50-pound bags of grain to 1-ton totes had been significant. Moving to this next level, the sense of progress was monumental.

Thor loaded the grain truck from the big storage bins at his farm, and I rode with him over the hills between Newfield and Enfield, feeling like a piece of popcorn as the truck harrumphed over the slopes. We stopped at the scales at a neighbor's farm to weigh the load. At the mill, Thor tipped the dump body up to feed the grain down into a plastic hopper attached to an auger. A sliding door in the back gate of the truck held in the grain.

"Are you ready for history?" Thor joked, and he opened the small door. The grain flowed smoothly at first, out of the truck and up to the bin, making smiles all around, but after a short time the auger came to an abrupt halt. Thor shut the little door, and Greg went inside to figure out an electrical problem.

"History always takes a while to happen," Thor said. "Why don't you have a seat?"

Lawn chairs with striped webbing sat under the grain bins, so I sat in the April sun. Benji Knorr, another worker, went inside to check on the milling, and Thor took a phone call in the driveway.

This is how my favorite flour gets made: slowly, and with lots of stops and starts. I think of flour as exciting because I'm so excited to know its route from farm to cupboard. Yet once I get to the mill, there's not a chorus line of stunning activity to observe. Neither grain nor millers do high kicks to impress me, and if I want a song, I have to sing it myself. Work is work, machinery is quirky, and each part of the operation is full of opportunities to fail. I love the way lively flour bubbles in the bowl as the pancake batter waits for the griddle to sizzle, but milling is far less dramatic.

Greg turns a switch and the grains start going up into the bin again. Thor opens the door of the dump truck. I take a picture of Greg, Benji, and Thor. The historic moment is preserved. The three of them look happy and patient. Ready for the next fail.

Greg and Neal keep tweaking operations. I wonder if the corrections will ever stop. There are four mills in the building: two Meadows stone mills that can run in series, a smallish Osttiroler that is used for cornmeal, and a hammer mill for buckwheat. The Meadows mills are set in metal housings, but the Osttiroler, an Austrian mill, is pretty as furniture, with pine planking and a built-in sifter. Before the new mill building turned two years old, it had already burst its buttons. An addition was soon built for more storage, with a spot for making and bagging cake and pancake mixes.

The physical space is not the only wall they face. Farmer Ground Flour can't take all the potential new customers who call. In the spring of 2014, as organic wheat prices climbed because of short supplies and high fuel costs, bakeries called out of the blue. Farmer Ground Flour's prices, which don't carry much of a surcharge for the bonus of being local, were suddenly competitive with national brands. However, the mill didn't have the capacity to meet the interest. Farmer Ground Flour makes 10,000 pounds of bread flour a month. New bakery customers need steady supplies. The mill had commitments to other customers, and had to say no.

Greg hopes that as the costs of commodity production rise, and Farmer Ground Flour's efficiencies increase, the cost of their flour will get closer to the price of organic products from far away. The price is already very fair for customers, because the goal is to make good food, not a boutique product.

"I'm waiting for the price of inputs, fuel, and transit to be enough to equalize the cost of regional production," Greg said. "As we scale up, our costs are going down."

Just as the community supported the bakery, the mill grew because people helped make it happen. Regional Access facilitated the logistics of delivery and dealt with the growing pains of a small business, letting Greg drop off bags of flour after hours, or coming to the mill when he couldn't get away. GreenStar, Ithaca's cooperative supermarket, has been the mill's best customer from day one. Erick Smith's operation, Cayuga Pure Organics, brought Farmer Ground Flour to markets in New York City and introduced countless commercial and home customers to organic, regionally grown stone-ground flour.

Greenmarket made more introductions. For instance, Finnish baker Simo Kuusisto found Farmer Ground Flour through June Russell. At the time, Simo was primarily working as a chef, and selling bread once a month at the New Amsterdam market. He was using the only rye he could find, from Guisto's, a mill in California. He was thrilled to meet Thor and Greg, see the fields, and develop a Finnish bread with deep New York roots.

"This bread is a staple in Finland, part of our Finnish identity," said Simo. Rye bread is very much a part of the Scandinavian diet, the first food people eat each day. His stand at Union Square draws many people who are loyal to 100 percent rye sourdough breads, either because of their heritage, or because of health interests. Rye has a lower glycemic index than wheat, and some people who can't tolerate wheat can eat rye. Simo started working full-time on the breads in 2013. The flat, round ryes he makes are rare in America, and he sends his Nordic Ruis all across the country. He sends Thor packages of bread and gives him Finnish rye seeds to grow. That kind of connection doesn't happen with commodity grains.

Corporate preachers say that business is all about relationships, but in most instances the scope of production segregates people by trade. Buy a bag of supermarket flour, and you can guess that the farmer, miller, and baker are strangers. Farmers who grow for the commodity market sell their grains by the truckload to grain brokers. These loads are tested for disease, weeds, and factors that affect the baking quality of flour. Mills and bakeries have formulas and specifications, and they don't need to talk to farmers about the numbers they need, or what these numbers mean. The numbers talk to numbers. There's no room or need for conversation. Millers and bakers seek the figures they want for their products. The cycle works without a lot of interpersonal communication.

When grains are grown and used in a blind, commodity process, they become anonymous except for the numbers that represent them. This is true for the malting, brewing, and distilling industries, too, as well as for baking.

Dial the scale down to a local level, and conversations have to happen from the ground up. The farmer has to listen to the soil, and pay attention to the weeds, pests, and diseases that accompany each crop. The miller needs to read every shipment of grain, both the figures that state its broad characteristics and what can be seen and smelled in a cupped handful of kernels. Bakers have to report what the flour says to them in the bowl.

The people involved with Farmer Ground Flour talk to one another often. They are reinventing a flour wheel, and their livelihoods are linked. As they go through their days, they have a lot of connections in person and over the phone. Greg knows what it takes to get the crop from the field. He and Thor discuss the agronomic aspects of different varieties, such as disease resistance, yield, and harvest ease. They talk about moisture levels, trying to find an optimal balance between what the grains need for storage and what works best for milling. After each harvest, they compare protein numbers on wheat to blend flours.

The bakery functions as a test kitchen. With the bakers at Wide Awake interpreting the flour, Greg, Neal, and Thor's jobs make more sense. The non-bakers learn what people need from the flour.

Industry standards at the commodity level short-circuit the need for such intersections. Large mills blend oceans of grain to create a more consistent product. Factory baking requires flour that will behave in the same way, batch after batch. Flour is a natural product, however, so conditioners and other additives help tame it and produce uniform vats of dough that can yield miles and miles of loaves.

Community-scale production has more room for grains and flour to say something else. To say, *Hey, this was a wet season, and the grain doesn't have as much protein as we'd like, so we have this flour this year, okay?* Grains articulate soil and climate. Farmers coax certain expressions by building nutrients before planting, and adding fertility at moments that will suit the final use of the crop. But the seasons have their say, and millers and bakers work within those articulations.

The collaboration among farmer, miller, and baker was obvious when I began visiting Ithaca in 2011. Over the years, I've seen how much receptivity is required within each profession. Everyone is working with variables, trying to charm natural systems to behave in predictable ways. This is in fact impossible, and it really hit me when I took a baking class with Stefan Senders.

"One of the great complaints about local flour is it's too variable," Stefan told the class, a group of home bakers and pros who came for a crash course in artisan breadmaking and the idiosyncrasies of local flour.

If the baker is not being precise in measuring the flour and water, adjusting for temperature and humidity, Stefan elaborated, a baker's choices introduce more variables than any flour ever milled.

Some of the students nodded. These skilled bakers were very aware of the small moves that can really impact dough.

Even flour from big mills varies from batch to batch. Because of the volume of grains handled, the five-pound bag of flour you buy in April might be from entirely different wheat than the same brand of flour bought in May. Day to day, a baker makes lots of tiny and inadvertent changes. Bad loaves come from inaccurate measuring, or lack of attention to the chemical and biological processes at play. Dough is a living thing, and the issues that affect its outcome are many. Blending flour to hit a predictable range of performance markers controls the variables somewhat, but only to a degree.

"Traditionally, it's said that the baker plays with time and temperature, but of course there's plenty more than that," Stefan told us. "You can adjust the quantity of flour or water. You can play with the length and intensity of the mix, the number of folds, the hydration of the dough, and so on."

Stefan was eager to convince us to engage with the process of making bread and be responsive. All day long, he recommended that we build a relationship with the dough. He may not have used those exact words, but that was the gist of it, and what his gestures suggested. Baking bread is a relationship. One that, perhaps because of my limited bread-baking skills, I'm somewhat reticent to enter into myself.

I've watched a lot of people work with dough. Puzzled over how bakers can know when a batch needs to be shaped, or needs more time fermenting. This intuitive sense and attention to timing has always stumped me, but Stefan's suggestion that baking bread is a relationship gave me an answer. The baker, the ingredients, and the fermentation process are in conversation.

The baker has to be alert to where the dialogue, or dough, is going. And local flour requires a little more listening.

Extra listening is essential when processes scale back from an industrial to a human level. All farming requires a certain amount of attention to nature, but when you can't rely on chemicals to control weeds and pests, farming asks for more. The collaboration may seem to begin with seed-to-soil contact, but that partnership between the farmer and the crop doesn't start in September, when Thor fits his fields (*fitting* is farmerspeak for preparing a field for planting) and puts wheat into his seed drill. The work began seasons before, as he built up the soil's tilth and fertility. His choices are linked to the beginning of farming more than ten thousand years ago. All those generations of observation and action are tucked inside the motions of planting each fall and spring. The farmer first has an exchange with the land, and then talks to the miller about what has grown. The miller in turn talks to the baker before the baker begins to chat with the dough.

Industrialized farming, mechanized baking, and the commodity system render these people and processes invisible. Breaking out of this system makes the processes transparent, and allows people to see each other and their respective work. Bread in a see-through bag on my counter says nothing of our collective social investment in grains and one another. Sliced bread hides the questions people have asked of seeds, soil, scythes, and mills. I want to see the interrogations. I want to know what answers best serve our personal health, and the health of the environment.

CHAPTER TWO

PUBLIC WORKS

I live near the Erie and Champlain Canals, which freeze early because they are shallow. I'm not a good skater, and my only grace is that I do not constantly fall. But I like to be on the ice, and canal ice is plain magic. I get to go under bridges and near barges and tugboats.

Above and around, life is being life. Cars, trucks, and buses pass in a steady stream. Houses sit still, waiting for people to come home. Yards hold promises of summer, with barbeque grills and picnic tables locked tight in blankets of snow.

Under me, the water is another self, solid and full of personality. Clear chunks let you see down and down through a champagne fizzle of air bubbles. On the surface, a whorl of maple seeds sits like a flower. Deep inside the ice, I can see the shadow of wheat.

Trucks, houses, and canals. Beer cans and ice skates. All of this stuff, the accumulation of extras that seem essential, our cell phones and convenience stores, the miracle of vegetable peelers that actually work; all of this stuff exists because of the stored energy of wheat.

Sometimes I look at what I'm holding and think, *That rolling pin is only here because a staple crop allowed us to focus on something other than survival.* We have grains to thank and curse for being able to drive cars and have jobs in offices. Figuring out a steady food supply allowed us to divert our attention to astronomy, indoor plumbing, and building skyscrapers. Grains fed us so we could feed our imaginations on other projects. Like the Erie Canal.

I know I see the world through grain-covered goggles, but wheat figures intimately in the history and function of the Erie Canal. One of the canal's earliest and most ardent promoters was a flour merchant. Once built, the Erie Canal shipped tons of grain and flour to New York City and the world. Grain carved a marvelous channel across the state, offering settlers in western New York a chance to earn a decent living farming or milling, and opening up territory beyond the Appalachian Mountains.

The Erie Canal is not the first time grains traveled in quantity. When stored at the right moisture levels, grain kernels are stable and can maintain their food value for years. Ancient Romans ate wheat from North Africa. Colonists shipped wheat and flour from the Hudson Valley to the West Indies. There is not a single moment when grains went *poof* in a puff of flour dust and left daily life. But the canal is a pivot point in our developing country, an era and an object that shows how flour production—something so central to the development of Western civilization—became disconnected from the experience and the minds of the vast majority of people.

This separation was a long process. Whole villages used to drop everything and help with the harvest. Think of European harvest paintings from the sixteenth and seventeenth centuries. Even van Gogh caught the wonder of people hand-cutting grain. But picturesque as it seems, that wonder was hard work. Harvesting grains invited ingenuity. Generations of people gave themselves to the project. The name McCormick is famous today because Cyrus McCormick perfected his father's reaping device. As farming mechanized, grains went from being a community event to being a commodity product; from something we saw and understood to something we use without much thought.

If I can't exactly map the slippery slope from community to commodity production, I can blame cheap transportation for aiding and abetting the exit of grains from daily life. The Erie Canal made shipping ridiculously cheap. When it opened in 1825, the price of bringing a ton of flour to New York City dropped from $100 to $10.

The thought of an inland waterway across the state first struck European explorers in the 1700s, but the idea didn't gain traction until the early 1800s. Flour merchant Jesse Hawley was the first person to promote the would-be canal.

Hawley and his business partner, Henry Corl, shipped flour and grain over water and land to New York City from the Finger Lakes region, paying up to $120 a ton. The freight traveled in wagons over primitive roads, and in boats on lakes, rivers, and a small canal. While shipping expenses may not have been their only business challenge, by 1806 Corl and Hawley faced bankruptcy. Each abandoned the enterprise and his partner.

Hawley fled to Pennsylvania to avoid debtors' prison. While on the lam, he started writing about a canal to connect Lake Erie with the Mohawk and Hudson Rivers. He sent a letter on this topic to the *Pittsburgh Commonwealth* in January 1807, and another letter to a judge in Buffalo. That summer, he returned to New York and began to serve a twenty-month term in Canandaigua, a small town near Geneva and Seneca Falls, the places where he'd bought grain and flour.

Debtors' prison was a common fate in the infant country. Canandaigua was too small to even have a jail. Hawley's imprisonment likely took place above a tavern, and probably involved him being shackled to the floor. Hawley had some liberty, and access to maps that helped him chart a course for the canal. If he didn't have maps then, he'd seen plenty, and knew the geography well from trying to find the best route to get his goods to New York.

Between October 1807 and March 1808, the *Genesee Messenger* published his fourteen essays about canals, and the particular merits of building a canal across New York. Many newspapers used letters to generate content for their pages, and it was common for people to write anonymously. Hawley wrote as "Hercules," hiding behind a pseudonym that indicated the effort the project would require. The first piece the *Genesee Messenger* ran was the letter he'd given the paper in Pennsylvania. The second one outlined a nearly spiritual imperative for the gigantic project:

> *It appears as if the Author of nature, in forming Lake Erie with its large head of waters into a reservoir, and his having formed this Limestone ridge into an inclined plane, had in prospect a large and valuable canal, connecting the Atlantic and the continental seas, to be completed at some period in the history of man, by his ingenuity and industry!*

The Author of nature, indeed! At the time, such flowery language was common, as was this kind of thinking. America's impressive natural resources were like marble waiting to be carved. The arrogance of the presumption, that man could finish God's handiwork, was excusing itself slightly by naming God as nature's Author.

Hawley had twenty months to sit still and think big. He never mentioned flour or grain in his writings. Perhaps he wanted to hide his identity and crime, or thought his arguments would be weakened if readers knew the writer had a commercial interest that would profit from the waterway. Grains, however, could not have been far from the mind of any reader, since bread was a significant food source, and none of that bread came in plastic sleeves.

In his letters, Jesse Hawley named the economic advantages that canals gave both farming and factories in Europe and England. He said a canal in New York was essential for utilizing the good farmland west of Albany. He devoted an entire essay to the economic advantages of shipping salt from Syracuse, and another one to potash. Potash was such a well-known and needed material that he didn't even list the many products it was used in—just detailed how exporting ash to Europe would pay for the canal. Trees needed felling anyway, to clear land for farming. Potash was used for fertilizer, leavening, soapmaking, and other things, in America and abroad.

While the thought of building a 350-mile canal with just mules and muscles seemed ridiculous, Hawley presented his case in an orderly and logical fashion. He calculated the reduced rates for shipping specific goods, and projected the entire cost of building the canal. He mapped a route, noting elevation drops and other geographic details. He proposed other canals, too, to help develop other parts of the country. While contemporaries thought his proposition outrageous, aside from the brash assertion that the Author of nature mandated the project his arguments today seem very rational.

I often wish I could be alive in another time and understand what things meant. Water especially intrigues me. I look at the Hudson River and think of PCB contamination. If the river has a purpose, it is recreation, not transportation. Intellectually, I know that the confluence of the Hudson and Mohawk Rivers made my area one birthplace of the Industrial Revolution, but I can't fathom how important water was to people at the beginning of the nineteenth century. Too much time and too many layers of technology insulate us from knowing the real value of water and waterways. But we can try to guess.

To Hawley, water represented both power and pathway. The flour he'd sold was milled at Seneca Falls, at the northern tip of Seneca Lake, one of the Finger Lakes. When the merchant could, he used water rather than roads, shipping on lakes as well as on the Genesee, Mohawk, and Hudson Rivers. The Mohawk was especially useful because a ladder of locks from a small canal, the Western Inland Navigation Company Locks, made it more navigable.

Hawley identified a transportation solution for his own business ventures and a script for nature that he knew could develop the economic engine of the country. This plan came from a deep familiarity with natural resources and financial survival that was not exceptional to him, but a part of figuring out how to live and be in the early nineteenth century. People read the world for opportunities, same as now. Hawley had moved from Connecticut to central New York because the land was better.

I like to think of him sitting with pen and ink, writing his way out of his confinement. Maybe the noise of the tavern below bubbled up and taunted him. Maybe he came up with sentences as he walked his "gaol limits" through the village, the sky his only shackle. Somehow his papers got to the office of the *Genesee Messenger*, whether he delivered them himself or someone took them from him. The printer set the words, letter by letter in lead type, locking paragraphs in place for inking and printing. How ephemeral that writing could have been! As forgotten as anything else. But Hawley's words did not evaporate.

The notion of a canal was gaining traction, and the writings of my flour savant reached legislators who were promoting a New York canal. When these advocates petitioned Thomas Jefferson, he rejected the idea. The president was not immune to the dizzying power and promise of the country. He had, after all, dispatched Lewis and Clark on their expedition to explore and report back on the natural resources and native peoples of the West. Still, Jefferson believed the canal was a hundred years premature, too big a dream to build.

If someone who dreamed so much for America doubted the canal, imagine how the project seemed to those who were not charged with the architecture of a brand-new country. Not everyone saw a divine invitation for the canal, let alone a civic mandate. Besides being expensive, such a grand enterprise seemed physically impossible. The young country had no engineers. All we had were people and animals and ideas.

At the time western New York was the frontier. Speculators and farmers were buying up the fertile land of the Genesee Valley, but even they did not want to pay for the transportation improvements that would allow them to better mine the territory's resources. Populations huddled near the Hudson

River and other waterways didn't want to pay for the canal, either, because they didn't believe it would help them.

Still, the project inched forward. Preliminary surveys were made, and a commission appointed in 1810, with DeWitt Clinton as its head. Like Hawley, Clinton wrote letters as he built his campaign, using the pen name Atticus. He targeted his sales pitches to different parts of the state, focusing on the benefits to each region. Since Clinton had no grain history to hide, he predicted that New York City would become granary to the world.

In early 1816, thirty separate petitions for the canal flooded the state legislature. Clinton's speech, called a "memorial," was widely circulated before and after he presented it to the legislature. This speech further articulated how the expensive surgery to the state's geography would develop farming and manufacturing not just in New York, but across the nation. He declared:

> *If it be important that the inhabitants of the same country should be bound together by a community of interests, and a reciprocation of benefits; that agriculture should find a sale for its productions; manufacturers a vent for their fabrics; and commerce a market for its commodities: it is your incumbent duty, to open, facilitate, and improve internal navigation.*

Two hundred years later, *commerce* and *commodity* have become negative terms, because they are usually aligned with corporate interests. But for DeWitt Clinton, Jesse Hawley, and others building America in the early 1800s, commerce and community were not at odds. The two were a unified pursuit, as they are in contemporary efforts to re-localize economies and food systems. Commerce meant, and means, people could exchange goods and services, and survive.

The word *commodity* has little use in modern community-building efforts, because it is shorthand for the distancing that's occurred between people and production. Commodities are anathema to people who are trying to restore the value of regional production of crops, food, and other goods. This linguistic discrepancy is telling of the differences between an industrializing nation and one that has offshored most of its manufacturing capacity.

When Clinton said that the commercial future of the nation relied upon developing this piece of infrastructure, there was an understanding that communities would grow around businesses. West of Albany, there were not many places white people lived, because they couldn't support themselves. The canal was not hard to finance because early Americans were anti-development, by any means. The scope of the project was just too large.

Chopping the work into tiny pieces solved this problem. The legislature passed a bill approving partial construction in 1816. The next year, DeWitt Clinton was elected governor, and construction began on the Fourth of July. At dawn, a cannon fired in Rome, New York, and dignitaries scooped the first shovelfuls of dirt. How lofty and beautiful, to let the sunrise underline this beginning. If only politicians still had such fanfare and style.

Once the canal was partially complete, wheat fever really took hold. Western New York had a mild case of the disease, caused by the reputation of its fertile soils. More and more New Englanders left their rocky farms to look for agricultural gold. The cost of shipping wheat and flour plummeted to $10 a ton. The little settlement of Rochester became the Flour City, harnessing the power of the Genesee River to mill 25,000 bushels of flour a day. Places like Wheatfield and Wheatville developed, tied in name and work to the new grain economy.

The waterway made revolutionary changes. While naysayers had feared that the canal would be a conduit for cheap English goods to America, the Erie Canal behaved just as Clinton predicted, fueling farming, industries, and trade. New York City overtook Philadelphia as the main port of the country, and wheat and flour were its prime exports.

Rochester's success added a new term to the language: *boomtown*. Mills there used both New York and Canadian grains, but this flour heyday only lasted about twenty-five years. The Erie Canal made Midwest farmlands more accessible, and processing shifted farther west. St. Louis, Missouri, assumed Rochester's milling crown. Rochester subtly renamed itself the Flower City, for a developing nursery trade.

The canal had dropped transportation costs and centralized milling. Gristmills moved from communities to the regions where more wheat grew. Flour production began to be hidden from most people's view, first in Rochester and then in the middle of the country. Does this mean that the Erie Canal was the villain that stole fresh flour?

Such questions were popular even in the early 1800s. The Industrial Revolution was not an easy pill to swallow, and people doubted the value of mechanization. Sylvester Graham lamented the loss of love and affection that homegrown wheat and hearth-baked loaves represented. (Graham is the man behind graham crackers, which were originally made as part of his prescription against a more personal social ill—masturbation.)

Graham was not the only one to doubt progress. Writers worried that advances in the kitchen were unfortunate. Chemical leaveners like baking soda and baking powder were especially suspect. Because they abbreviated work, did they also abbreviate the love a mother could feed her family?

As I romance fresh flour, I am aware that I might be lobbing a guilt ball, too. Touring bread and grain projects, I've wondered what level of production is correct. Shouldn't flour and bread be made close enough to home that you can see the fields and know your farmer? Shouldn't you meet your miller and baker, too?

Or am I depriving my family of meaningful connections to the land and to one another if we don't grow all our grains? Should I set up a bicycle mill for the boys to grind our flour? Am I a lousy mother if I don't make sourdough bread? Maybe the only righteous bread comes from a village baker who uses local flour.

Let me collapse the telescope of these questions into one concern. Shouldn't we all have fresh flour? My answer is, unequivocally, yes.

I have no ambivalence about this. The moment I first tasted fresh flour is fixed in my head. In early August 2010, my husband came home from a trip to Connecticut and brought me a cookie. Jack is an arborist, and he had been climbing trees at fancy homes, installing light fixtures for a landscape lighting designer. Jack had slept in hotels while I managed kids who badly missed him. You get a hotel and I get a cookie? Gee, thanks.

I should not have been so ungrateful. That single cookie opened an entire world for me. The oats and wheat screamed for attention, even though the chocolate and butter had plenty of flavor, too. Jack handed me a pamphlet telling the story of the Wild Hive Bakery and its New York–grown and –milled grains. I was thrilled to learn people were making flour nearby.

I had been on the prowl for flour. In the spring I'd taken a walk at the Poestenkill Gorge, carefully picking my way down a steep path to the rocky edges of the stream. There I spied a grist stone, left who knows how long before. The Poestenkill had powered woolen mills and mills to grind slate into powder for paint, to make the color known as battleship gray. Finding the flour millstone should not have surprised me, but it did, and it got me curious about when and why my city quit making flour. My investigation began when I took a bite of that oatmeal ganache cookie bar.

Baker and miller Don Lewis put this food in my husband's path. If Jesse Hawley helped open the door to anonymous flour, Don opened the door to flour of known origin.

Don is a pioneer in grains. For years he sold baked goods with fresh-milled flour at Union Square Greenmarket and other farmers markets in the city. When Jack found that cookie, Don was running a bakery not far from the Taconic Parkway, an old divided highway that's so narrow it doesn't allow trucks. Clinton Corners is in Dutchess County, just north of New York City. Around here, the houses and yards are well tended, and so are the fields. Horse fences can be architectural marvels. The bakery closed in the fall of 2012, but a few miles farther from the highway Don now mills in a barn that almost became a house. The building still looks like an old dairy. The inside is hand-somely renovated, and its marble fixtures and hardwood floors are covered in flour dust. The juxtaposition of the building's rustic exterior and frilly/milly interior always strikes me as a little odd. One summer day, I came to hear Don's flour story from start to finish. I knew parts of it, but I wanted the whole.

We sat in his office, a small room with a window overseeing the gravel parking lot. A fan fluttered papers on his desk. Don took off his cap. He often wears a hat with a small brim, and he always has a mustache. This modest facial hair reflects his reserve. A bushy gregarious stash would seem out of place on this man, who holds his cards close to his vest. I've known him for four years, but he remains something of an enigma. I was happy to get the details of his story, and try to understand what led him to fresh flour.

I already knew the basics. Don ran into local flour in the mid-1990s thanks to another grains pioneer, farmer Alton Earnhart.

"I heard about this person producing organic animal feed and I ran up there," Don said of his first trip to Lightning Tree Farm. Don wanted to feed the chickens he kept, but Alton had other grains, too—5,000 pounds of wheat, his first harvest. He offered Don a cookie made with the wheat he'd grown on a dare. When other people said he couldn't raise wheat, he tried and succeeded.

"Alton gave me a bag of flour and said, 'Well you're a baker, maybe you can do something with this.' I stuck my hand in it, and I knew right away things were going to be different," Don said.

All flour feels nice to a baker, but this kind felt and smelled very good. Whole wheat, its freshness leapt at his senses. Before long, Don figured out how to work with the flour and incorporate it into the honey-based baked goods he made and sold at farmers markets.

Adding homemade flour was a natural expansion, one that echoed the way Don grew up. His father ran a chicken farm and kept bees. When his family opened a restaurant, everything was made from scratch. There was no Boar's Head for the deli sandwiches, just freshly roasted beef. His father's specialty was barbecued chickens. The restaurant was stocked with food from the area, but *local* at the time was just an adjective used for sweet corn. Although fresh flour was new to Don, he knew this kind of food.

"This was the way it used to be. Simple," he said. "Flavor, texture, nutritional density."

Don added the new ingredient to everything from baklava to bread. Instead of making a single item that highlighted the flour Alton milled, he wanted a part of each product to be made with fresh flour. First 8 percent of the recipes had fresh flour, then 12 percent, then 15 percent. The percentage grew every year, and so did the acreage Alton planted. Other New York grain farmers provided seeds and offered crucial post-harvest handling lessons. Soon Alton couldn't keep up with the milling, so Don set up a mill. The power company was slow in getting electricity to his barn, so he worked in a trailer in his front yard, running an extension cord from his house.

"I milled some really terrible flour that first winter, but I figured out how to make it work," Don said.

This gave him two educations at once. He learned how to mill and how to compensate bake—how to adjust his recipes for the ways the flour behaved, changing amounts of liquids, for instance, and bake times. The steep learning curve burned out a 12-inch set of stones from Meadows Mills in North Carolina, which he traded in for a 20-inch mill. As his milling improved, his wheat needs increased. Farmers Pete Gianforte and Klaas Martens started supplying him. All the while, Don kept increasing the percentage of fresh flour he was using in his baked goods.

"I thought it was important to get it into people's stomachs. In order to understand it they had to digest it. And really digest it," Don said. The food would talk to their bodies and their brains, reinforcing the new concept.

"I didn't want it to be a seasonal product like strawberries used to be, and have people say, *Oh, this is wheat season?*" Don said. "I didn't want to break that continuity. The positive effects of eating freshly milled flour are maintained even if it's a percentage of the flour. Your taste buds are attuned to it. Your digestion's attuned to it. That good feeling you get from eating it stays with you."

Eventually he was able to run the bakery entirely on what he milled himself: 20,000 pounds of flour. This was a milestone.

Along the way, the ingredient took on a job. This food was information, part of a mission to provide a supply of regional flour. At some point, Don changed the name of his business from Wild Hive Apiary to Wild Hive Farm Community Grains Project. Earlier, before he'd started baking, he had a commercial beekeeping service. His bees pollinated orchards along a 50-mile stretch of the Hudson River. Now his food was pollinating people, and an idea he wanted to manifest.

"This is what people should have access to," he said of the flour. "We should be able to buy this. We should be able to buy products made with this, and I resent the companies that have taken it away and are trying to make it so that you can't access that quality. There's no reason for that. Not in this country."

Don's example has inspired many other grain projects. He was the keynote speaker at the first Kneading Conference in 2008, and his model encouraged conference organizer Amber Lambke and baker Michael Scholz to start up a mill in Skowhegan, Maine. Now the two run Maine Grains at the Somerset Grist Mill.

Lightning Tree farmer Alton Earnhart, as well as other farmers, demonstrated that local grains could be grown, and Wild Hive miller/baker Don Lewis showed that local flour could make bread. These two lessons helped Greenmarket, the organization that runs farmers markets in New York City, mandate a rule for bakers to use regionally produced flour.

Greenmarket is part of GrowNYC, a nonprofit whose mission and work is environmental in scope. The Office of the Mayor is a sponsor of the umbrella organization, providing in-kind support through housing the organization, but GrowNYC does all of its own fund-raising and its projects are independently governed. Greenmarket has a mission to support regional agriculture and bring fresh food to the city, through farmers markets and other programs.

Greenmarket runs 54 farmers markets, and vendors can sell only what they make or grow. The first market was in Union Square in 1976, and the definition and enforcement of that producer-only rule has morphed over the years. The use of regional flour in baked goods became a requirement in 2010.

"Conversations about flour began in 2000, 2001," June Russell said in an interview at the Greenmarket offices in the spring of 2014. June is the farm inspector, and also works on the Greenmarket Regional Grains Project. We were sitting in offices at City Hall, with the organization's director, Michael Hurwitz, and Heidi Dolnick, who helps on many grain projects.

The room felt staunchly institutional. Thick, tall walls were painted bright white and an unfashionable light green, suggesting bureaucratic inflexibility, but our conversation about grains reminded me how the walls between Greenmarket and the food world are actually very thin.

Chefs and consumers come to Greenmarket for locally grown and produced foods. The organization is a gateway to fresh food in the city, and a gatekeeper of what that food is and where it comes from. Greenmarket is a powerful driver of regional agriculture. In 2007, the push for flour production began.

That year, Michael became director, and June, who had been working as a market manager, began overseeing farm inspections. As new blood, they knew they had to do something about the bakers. They didn't have to meet the same criteria as other vendors, which was unfair. Plus, many baked goods were inferior in quality compared with everything else that was for sale at the markets.

"In a producer-only marketplace, the baked goods stood out," Michael said. While made by the vendors, the majority of ingredients came from far away. Don Lewis was unique in that his honey and flour were local.

"When the market started, space was not at a premium," June explained. Baked goods were included so shoppers could fill their baskets and not have a reason to shop elsewhere. Thirty years later, good bread was plentiful in and around the markets, and many customers were aware of what foods came from where. The time was ripe for change.

June had heard complaints about the discrepancy as a market manager. Farmers had to comply with regulations, and felt that bakers should not be exempt. What did banana bread made from commodity flour have to do with regional farms? The issue percolated to the surface now and again, but nothing got resolved. A 2004 bakers meeting prompted a rule encouraging the use of regional flour.

"The rule was unenforceable," Michael said, because it was written with a big fat *if* escape clause. If regional flour was unavailable—because of crop limits or a mill fire, for instance—alternative flours were okay.

"We really started to think about how bakers could help drive this mission to support regional agriculture," Michael said. The first conversations focused on the criteria they might use to judge bakers, in terms of the taste, quality, and freshness of their products, as well as their use of regional ingredients, including but not limited to grains. June and Michael knew the market well enough to move slowly, taking the pulse of bakers at meetings, and surveying the marketplace to see what was available.

While doing farm inspections, June kept an eye out for flour and other foods that were not at the market. Staple crops were lacking throughout the region. No one grew dry beans, and only a few mills handled grains grown in the Northeast. The irony is that mandating the use of local flour could actually stimulate production. A rule intended for bakers would ripple back to producers and stimulate grain growing and milling. In other words, demanding a supply would increase that supply, but you couldn't legislate the use of something that wasn't really there. Just because bakers were functioning outside the market's mission didn't mean they could be thrown under the bus.

"We spent two years talking about changes in our rules for bakers," June said. "There was a lot of resistance, and we listened to that resistance. It gave us a lot of lead time for the implementation."

Many Greenmarket bakers said that local flour, when available, didn't work well. Some bakers, though, were more receptive. Don Lewis was proving that local flour did work, and others were ready to get behind the idea of change. Dan Leader from Bread Alone had helped introduce European hearth-style baking to America. Matt Funicello from Rock Hill Bakehouse had used New York State flours in the early days of Champlain Valley Milling in the 1980s, and was eager to use them again.

As the organization discussed what they could expect of bakers in terms of usage, June kept looking for sources. She found Daisy Flour, a Pennsylvania mill, at a conference put on by PASA, the Pennsylvania Association for Sustainable Agriculture. She also met Eli Rogosa from the Heritage Grain Conservancy at a meeting of the Northern Grain Growers Association in Vermont. Eli was working on the Northeast Organic Wheat Project, a USDA SARE grant exploring heritage grain growing.

Shortly thereafter, June met Elizabeth Dyck, an agronomist who was working with NOFA-NY on projects with farmers growing grains and potatoes. Elizabeth fast became an ally of Greenmarket's grain work, and came to a meeting of Greenmarket bakers in April 2008. She brought samples of several wheat varieties. The update on local flour at the meeting was this: "The good news is much is percolating; the bad news is consistent availability is a long way off."

On her farm inspection rounds, June had found Erick Smith, and urged him to start the mill. The prospects for this mill, which would become Farmer Ground Flour, were hopeful, but it didn't help set a new rule. Before that could happen, Greenmarket had to bolster supplies. Daisy Flour and Champlain Valley Milling had the capacity to mill more local grains, but sourcing food-grade wheat in the Northeast was tricky. Progress was slow, but happening. And then a landmark story helped make more happen.

Greenmarket is well aware of the power of storytelling, and fed news of the burgeoning grain movement to *New York Times* reporter Indrani Sen. Sen wrote a story titled "Flour That Has the Flavor of Home." The article featured Don Lewis and Alton Earnhart, Mary-Howell and Klaas Martens from Lakeview Organic, Jack and Anne Lazor from Butterworks Farm in Vermont, and Eli Rogosa.

Stories put pictures and possibilities in our minds, and help broker change. This *New York Times* article about grains leveraged change, too, announcing nationally that grains could be grown outside the traditional wheat belts. The story is about how the owners of the building where Don Lewis now mills found him, and he found room to expand his milling operation. In the Finger Lakes, Farmer Ground Flour saw the *Times* article and felt more confident that the crazy thing they were doing would succeed.

The article also strengthened connections between renegade growers and millers. Jennifer Lapidus, who had used tobacco money to put together a mill in North Carolina, contacted Greenmarket. Her Carolina Ground flour would never reach New York bakers, but her experiences working outside the commodity grid would serve other small grain projects.

The *New York Times* shone a light on the revolutionary prospect of regional grains, generating awareness in people who had never heard of the idea. The newspaper showed that new thoughts about grain were possible, which made other new thoughts about grain possible.

At Greenmarket, the story made the push for local grains feel more solid. June kept gathering resources and working with others to figure out when to implement a rule. At a bakers meeting in 2009, they announced that in 2010, bakers would have to use at least 15 percent regional flour in their products. Mill representatives came to the meeting, introducing their goods. Roberta Strickler from Daisy Flour came, and so did Greg Mol, toting 10-pound bags of Farmer Ground Flour. A handful of bakers took the flour and got a jump on incorporating it into their production.

Bobolink, a cheesemaker that also made bread, and Orwasher's, an old bakery on the Upper East Side that was developing an artisan line, became immediate customers of Farmer Ground Flour. Rock Hill Bakehouse started working with North Country Farms, a stone mill near the Canadian border. These and other early adopters of the Greenmarket rule helped develop demand, and develop distribution too.

Since instituting the rule, the grain industry in the region has blossomed. Providing markets has been a boon for mills. Older ones have been given new life, newer mills are growing, and new micro-milling operations are up and running.

Farmer Ground Flour and Small Valley Milling built new facilities. Wild Hive is contracting with other mills in the region. Maine Grains is selling oats to New York City bakers. Castle Valley Mill and Weatherbury Farm are coming online in Pennsylvania. Heritage wheat breeder Eli Rogosa is working with farmer Klaas Martens, who is growing out heritage varieties like Red Fife wheat, and Champlain Valley Milling is turning those grains into flour.

Champlain Valley Milling and Daisy Mill have expanded operations to help meet the demand. Interest in regional grains and flour has spread beyond the farmers markets, too. As a prime purveyor of local foods in the city, chefs turn to Greenmarket to source local ingredients. The grains project arranges introductions to grains at farmers markets by stepping into vending mode to run a retail stand selling dry beans, grains, and flours from the region. Weekly at the Union Square Market, and monthly at other markets, people can sample dishes and buy flours and whole grains like spelt and emmer from Small Valley Milling or Lakeview Organic Grain, or oats from Maine Grains. The project hosts special events like an evening emmer tasting, bringing speakers to discuss the food's history and use.

The rule for bakers went into effect in 2010. After a year, through reviewing invoices, Greenmarket traced the use of 60,000 pounds of regional flour a month. June works with partner organizations on grants to help create more opportunities for grain growers and small processors in malting, baking, and brewing. The project has a website that is a tool kit for people who want to get involved at any level, from just shopping for regional grains to starting a malthouse. Case studies and filmed interviews with model producers like Valley Malt in western Massachusetts and Wide Awake Bakery provide technical support. A map of the grain system they've helped to create in the Northeast is here, too, and the network of farms and mills is impressive.

Don Lewis is still growing that system, but not through farmers markets. He has stopped baking and focused on milling for the past few years. Wild Hive produces 100,000 tons of flour a year for Eataly, the New York City Italian food emporium. He's helping the Chicago branch of Eataly to produce regional flour, too. Eataly is the largest single consumer of regional grain products from the Northeast. Nobody can buy that transformative oatmeal ganache bar that first opened my eyes to the flavor and potential for local flour, but Don continues to find ways to introduce grains to new people and markets.

We eat what we can get. Food systems are about infrastructure and markets, whether at the national, international, or regional level. The Erie Canal was a huge piece of infrastructure that determined how food would be grown and delivered, and it precipitated a cascade of sweeping alterations to the way we get our daily bread.

The Greenmarket Regional Grains Project can't orchestrate change of the magnitude that the canal facilitated. Almost two hundred years of private and public investments have built the framework that sets prices for food and parameters for farming. Railroads, the Eisenhower Interstate Highway System, farm subsidies, fuel subsidies, international trade agreements, international markets, and shipping routes stack the deck of agriculture in favor of corporate agribusiness and large-scale production.

Greenmarket moves small mountains against one of the big-daddy commodities, brokering relationships and laying the groundwork to revise grain production and use. By building markets, and creating demand, they are developing farmers, milling capacity, and distributors—the necessary infrastructure for a modern-day regional grain system.

More grains are in the ground, and more flour is at the mills. People are making connections to the fields and farmers that make their food. Local bread may never be on every table, but the agricultural drift that the Erie Canal began is shifting, thanks to enterprising farmers like Alton Earnhart, bakers like Don Lewis, and the Greenmarket Regional Grains Project. These people and public works have shuffled our connections to staple crops. They are engineering everyday revolutions.

CHAPTER THREE

PEOPLE AND PLANTS

I grew up in the country without ever noticing that land made food. My mother planted perennial gardens because she loved the way the plants looked, and the work felt good. My father grew more stories than food. Any vegetables that came to the table seemed incidental to his tales of critter combat. When I rode my bike past cows and corn, trees and hay, I felt connected to the road. I was slipping through the edges of nature, and ignored the center.

Later, my own family gardened, and together we tended the harvest. My son and husband studied seed catalogs and discussed the names and aspects of plants. For a while, I tried to join the conversation, but I couldn't keep my mind on plant names, common or Latin. The information hit my ears only as sounds, not words, like when adults talk in Peanuts cartoons, saying only *Mwa, Mwa-Mwa*, or *Mah, Maah, Mah*. I love to cook, and I love people who love plants, but my brain is immune to botany.

This was apparent the first time I saw the Plantations, Cornell's unfortunately named demonstration gardens. Jack and I brought the kids to tour the plantings. We walked around the inside and outside of a square-shaped garden. Each plant was tagged, and Jack and Francis studied the peppers. Francis, who was thirteen, wrote down the names of the ones he admired. I looked at the rosemary and lavender, and felt wilted in the late-afternoon heat. I wished I could pay attention. I wanted to connect with my family, and as a food writer I felt obliged to know more about my subject. I couldn't force it, though. When I told stories of how food got from dirt to dinner, my engagement was all about the people.

Jack and Francis were intent on what they saw, so Felix and I wandered. He stuck his hands in a water fountain. We crossed the road to the heritage gardens, which showed four different periods in American food gardening. One focused on plants from the Gilded Age. A tomato plant was tagged with a note about how the desired size of the fruit changed as the canning industry developed. I wanted to know more about the cooperation between people and plants, and how that influenced what we eat.

The work that goes into food is invisible. A farmer's curiosity about seeds and seasons, a tomato grower's desire to better squeeze the harvest into a can, disappears. Worse yet, the history of a food grows maligned. The positive ingenuity that benefited the canning industry in the late 1800s gets lumped into the twentieth-century production of cellophane-sturdy supermarket tomatoes. So much good tinkering has been done to advance the food supply. And so much more tinkering is required as food production scales down from an industrial level.

Standing in the garden at Cornell, I could imagine a person working to make tomatoes more uniform in size. A real, particular, flesh-and-bones person, not the corporate "personhood" that eaters of a certain stripe vilify as we rail against the ills of the Food Industry. Since our current food system is a failure, we believe everything that led to it must have been malicious or somehow suspect, full of selfish motives. Today there is a deep suspicion of science and plant breeding. The tag in the garden gave me a point of entry, and I was primed for more information.

I began to get a sense of the human muscle that is flexing to make food. People—many, many people—are putting their shoulders to the work of feeding us. A few months later, in that same place, I got another window on that work.

I was meeting with plant researcher Julie Dawson outside the visitor center at the Plantations.

"People don't think about everything that goes into getting wheat in the field," Julie said. "We're benefiting from ten thousand years of agriculture."

I nodded, as if I understood farming, or wheat. We were sitting at a table outside the welcome center. The fall day had warmed up, and I squinted at the sun shining on her glasses. I was just meeting her, and only barely acquainted with grain production. I felt keenly aware of how little I knew, but luckily my ignorance didn't matter. Julie answered my vague questions with a dictionary-like assurance, telescoping onto the broad world of plants.

I was wondering about the concept of bread building community. In small-scale grain projects, were people collaborating more than in other farming styles?

"From what I saw in France, I would say it really does connect people," she said. "Farmers have a strong community built around seeds and varieties."

Julie had been working with farmers to develop new wheats, seeking good baking and growing qualities for farmers who milled their own flour for baking, or worked closely with millers. The national agriculture research institute of France, the Institut National de la Recherche Agronomique (INRA), was leading the investigation.

Physically, Julie is thin, and temperamentally she is a very focused person. Her curly red hair is short and tight on her head. She got her PhD in wheat breeding at Washington State University's (WSU) Pullman campus, testing wheat in organic systems and developing better methods to select varieties that would do well under organic systems. As part of her research, she set up roundtable meetings with organic and conventional farmers—just got them together and heard their concerns.

While it might seem a necessity for plant breeders to be in touch with the people who use seeds, most breeding today is done for industrial agriculture. Such farmer–breeder contact is not critical, because the scope of plant breeding is set by industry parameters. The paradigm at WSU is very farmer-forward, however, and most wheat grown in the state is still developed by the public university program. Working directly with farmers made her work very relevant, and also fun. The experience made her seek opportunities like the one in France, where farmers are partners in the research process, not just users of the work that's generated.

That day at Cornell, her respect for farmers was clear, and it helped me see a bigger picture of the long human project of people and plants.

"Farmers have historically been the stewards of natural resources and genetic diversity," she said. "Farmers do more than provide raw materials. They are actively conserving this collective resource."

For millennia, generations of people and generations of plants have been collaborating. She gave me a sense of the back-and-forths of the relationship, the investments that people put into plants, and the investments that plants put into our heads as we have carried seeds forward through time. I could see a chain of people whispering in each other's ears, passing seeds like notes. Seeds were things people protected, used, and cherished.

"I worked with farmers actively selecting populations of wheat, landraces, and historic varieties," she said. "We crossed landraces and other historic varieties to create populations that had more genetic diversity."

She gave me a lesson in seed categorization. Modern seed varieties come from plant breeding programs, and grow plants with very predictable traits. Historic varieties also come from breeding programs, but date before 1950; modern varieties have usually been developed for farm systems that rely on synthetic fertilizers and pesticides. *Heritage* and *heirloom* are other names for historic types that come from human selection, but predate the existence of professional breeding programs.

Landrace seeds exist for almost all crop species, except for very new or industrial crops. They differ from modern and historic seeds in that farmers, not seed breeders, have developed them. These varieties have been cultivated in traditional agricultural systems, and are well suited to the areas where they were grown.

In the United States and Europe, most landraces were lost when modern varieties became widespread after World War II. Some landraces have been preserved in gene banks. Some have been preserved by farmers who never stopped growing them, and this is better because the seeds have continued to evolve and adapt. In a gene bank, seeds are protected, but they cannot adapt.

"If you don't have genetic diversity, you can't evolve," Julie told me. "Any one year presents stress, and, over time, their descendants are more able to deal with stress."

The only way for that to happen is to have enough individual plants in the field to respond to the pressures of climate and environment, and pass on their changes to the next generation. All of the evolutionary benefits the plant developed go static when they are stored. Think of how we trap the wisdom of our elders in nursing homes, and squander the accumulated life lessons of entire generations.

Working with the French farmer-bakers, Julie and other researchers were making crosses between landrace and historic varieties, and planting the crosses on farms. Farmers observed the plants growing and selected types that did well. This built on work that farmers had undertaken on their own; organic farmers started using landraces and historic varieties because these performed better than the modern varieties.

This is an example of participatory plant breeding, a union of the best of traditional and modern approaches that puts the tools of science in real farm circumstances. Julie has a long history with this style of research, starting in high school. She interned with researchers at Cornell who were involved with participatory dry bean breeding in Honduras. She was interested in organic agriculture and plant breeding, and participatory research was a way of valuing both the farmer's knowledge and the researcher's knowledge.

The work in France was an especially strong collaboration because it grew directly from farmers' interests, expanding on-farm inquiries into the laboratory, and bringing lab results back to farmers for further on-farm plant selection. Evaluation of the results sometimes happened right on the farm, too; some farmers tested grains by milling them into flour and baking bread, making a tight, direct link between how the grain grew in the field and how it baked. Some of the farmers made pasta as well; and others were linked to small mills.

Julie's work was part of an effort to help farmers work inside a highly regulated seed environment. In Canada and Europe, seed varieties must be registered in a catalog. This is a costly process. Consumers are enchanted with older seed varieties, but farmers have limited chances to reach this market because these seeds are not in the catalog. The project also sought to develop seeds for organic systems; not many registered varieties suitable for organic and low-input agriculture are available to farmers.

The work was not strictly about grains, though the farmers involved with grains were very active. That is because there is only one variety of wheat developed for low-input systems available in the seed catalog; 80 percent of the organic wheat acreage in France is planted with that type.

The farmers she worked with were part of Reseau Semences Paysannes, the Peasant Seeds Network, a collaborative effort to address crop biodiversity within the limits of this commercial seed environment. The farms she worked with are very diversified. Farmers might have a mill and a wood-fired oven, and also press oilseeds. Multiple crops benefit the health and diversity of the farm, and multiple products like honey, oil, bread, and pasta help the farm's economic health. Julie stays in touch with people involved in the projects, keeping abreast of how the seed work continues.

Julie now works in research and extension at the University of Wisconsin, but she stays in touch with projects in the Northeast as well. She is a quiet, compelling force on a number of grains projects, and a strong intermediary lobbying for the health of farms and farmers. Julie Dawson is a part of that continuum she helped me see, the long line of people sharing information about plants.

Elizabeth Dyck is another person in that line. The founder of OGRIN, the Organic Growers Research and Information Sharing Network, she lives in Bainbridge, New York, about an hour east and south of Ithaca. She travels across New York and Pennsylvania connecting farmers to one another and to practices that can change how and what they grow. She is a champion of both crop rotations and small grains.

I met Elizabeth the same day I met Thor Oechsner, in January 2011, at NOFA-NY's winter conference. (The acronym stands for the Northeast Organic Farming Association of New York.) The two of them were giving a talk on ancient grains, and they lit up the beige room much brighter than its drowsy lighting. As people sat down, Thor stood at a table, rows of little ziplock bags lined up in front of him.

"Ancient grains! Get your ancient grains!" he chanted like a guy selling peanuts at a stadium. He tossed bags of emmer to people he knew. Elizabeth was nowhere near as goofy, but equally well known to the crowd, and just as engaging. They were sharing a ninety-minute presentation, but she let Thor speak for an hour. Over the years, I've learned that this is characteristic of her. Anytime she sees the chance to let farmers talk, she steps aside. This inclination seems native to her midwestern roots and her family's tradition in farming.

"I've been growing vegetables organically since I was a kid. I started doing that with my father and never went beyond that to any of the so-called modern practices," Elizabeth said. Her father was a Mennonite who immigrated to Canada from Ukraine after the Russian revolution. She spent her early years in northern Minnesota then went to high school and college in Pennsylvania, where her parents taught at a small state university.

Elizabeth wears sneakers and practical work clothes. Summers, she wears a tan hat that keeps off some sun, but nothing can trap her own energy. Like her gray hair, which spills out from a loose bun, she and her ideas are on the move. Her devotion to her work inspires the same from farmers, who dedicate land and time to her projects.

One hot and humid day I watched her walk a field, broadcast-seeding rye. Thor had just planted the field with buckwheat using a tractor and a seed drill. The plots for her experiment were too small to use the seed drill, so she filled her bag twice and covered a lot of ground, regardless of the high-noon wilt. Her face was red, and she limped a little; she drives so much that she has difficulty with one leg. She didn't complain about the weather or any pain she felt. She just planted the seed, eager to see whether this old-fashioned companion-planting style might boost the yield of the rye crop.

Elizabeth's research career began in the late 1980s at the University of Maine's Sustainable Agriculture program. Studying weed ecology, her interest in organic vegetables led her to small grains, because organic systems control weeds through crop rotations. Planting grains and cover crops is essential in the non-chemical management of plant diseases and weeds. They build organic matter in the soil, and provide a break from the pest and weed threats that are created by repeatedly planting the same area year after year to crops from the same plant family.

In her first job after graduate school she put this research into practice. In the early 1990s she worked with Kenyan farmers, trying to get legumes back into crop rotation systems. These valuable protein-rich crops had fallen out of use because of the Green Revolution, which applied high-input agriculture to developing countries. The International Monetary Fund and donor countries had subsidized chemical farming for a long time, dramatically changing the way people grew food.

"Just as I got to Africa, those subsidies were stopped, so farmers were left with crops that needed very high levels of inputs, both fertilizers and pesticides, and no money to pay for them," said Elizabeth.

Legumes feed the soil, by capturing or "fixing" nitrogen from the air on their roots and building fertility; and they also feed people, as a vital source of dietary protein. Her work took place in formal scientific settings, and also on farms at twenty locations throughout Kenya. Prior to the Green Revolution, legumes had been intercropped or used in rotations. But pulse crops like green beans, field beans, pigeon peas, or chickpeas were gone. Gone, as in not for sale or stowed away in seed banks, as they might have been elsewhere. So Elizabeth had to scrounge for seed.

While the United States has strict rules about bringing seeds into this country, there's little regulation of the flow of seeds out, and African countries have limited protocols to keep seed supplies safe. This worked to Elizabeth's advantage as she sought beans, peas, and forage legumes, some of them tropical varieties. Her mother was a key player in this effort, helping her import seeds from California and Florida, and a few rare types from Europe.

At her next job, she also had the dual tasks of identifying seeds and building markets. She helped develop the Elwell Agroecology Farm, a University of Minnesota research farm, as an organic resource.

"There were very few organic farmers in that part of the world, so we were involved in a lot of conversion projects," said Elizabeth. This was in the late 1990s. Farmers interested in organic production in the corn-soy belt needed to get to know and use alfalfa and small grains on their farms and find ways to market these secondary crops.

While she worked on her doctorate, the importance of sustainable rotations became apparent in her experiments and the literature she read. Once she was working with farmers, she saw that the ecology she understood had to be economically justified.

"You have to help the farmers develop or reach markets; otherwise they can't possibly grow them," she said. "Farmers the world over have to grow crops that are profitable. This seems like an obvious thing, but it can be an Achilles' heel of research projects. *Oh this is great, does such and such for the soil, such and such for this and that*, but what does it do for the farmer's pocketbook?"

Collaborative work with farmers, she believes, is really the way forward. Rather than researchers setting the content and objectives, she favors a participatory approach in which farmers are front and center in the decision-making process and the development of research projects. This is why she founded OGRIN and established herself as an independent researcher. Farmers set the agenda for the work she pursues.

While independent, OGRIN is not an island. Elizabeth works with Cornell, Penn State, NOFA-NY, the Pennsylvania Association for Sustainable Agriculture (PASA), North Dakota State, and the state universities of Vermont and Maine. Greenmarket Regional Grains Project is a strong teammate, too.

These groups work on joint projects, seeking funding to support farmers growing food-quality grains in the Northeast. North Dakota figures in the work because of emmer. A few German farmers brought emmer to this country. The seed probably came from Russia, and the American descendants of these immigrants never stopped using emmer and saving the seeds. The university is now shepherding seeds through research trials to develop more types suited to the growing interest in ancient grains.

The work is wide ranging, from trials to see if the protein in grains increases according to different schedules of nitrogen application, to classes at Wide Awake Bakery. Stefan Senders teaches one-day immersions in bread baking. These have allowed farmers to incorporate baking into their operations, which is exactly the result Elizabeth wanted to see. Using small mills and different kinds of ovens, a few grain growers are now capturing all the value possible from their crops, making bread to sell at farmers markets or farm stands.

One-day courses are manageable for busy people. Elizabeth seeks to create experiences that are relevant to farmers and others involved in small grain enterprises. Partnering with PASA and NOFA-NY, OGRIN hosted a seed cleaning workshop at Ernst Conservation Seeds. The goal was not to produce more seed companies, but to expand knowledge about the general task of seed handling. This information is directly applicable to the job of producing high-quality grains. Touring the Pennsylvania seed company, people saw processing and storage equipment and methods, taking home hints to apply to their own projects.

This workshop was part of a grant on value-added grain projects sponsored by USDA OREI (Organic Research and Education Initiative). Another workshop brought farmers together to hear from a Michigan miller who runs a historic windmill, as well as from a Massachusetts maltster. Participants got a quick but intensive look at the work of malting and milling. There was enough information for farmers to see what these businesses need from them, and for people considering start-up mills or malthouses to better understand the ventures.

Such investigations explore both farming and marketing, the elements Elizabeth knows are critical to supporting farmers. I love the work she does for grains, but for her the undertaking is a means to an end. Building markets for grains is essential for sustainable agriculture.

"I've studied the history of rotation and cropping systems, and I know without a doubt that if cropping systems don't include small grains, they will fail," she said. "All cropping systems may fail eventually, but they'll fail a lot faster if they don't include small grains."

Organic agriculture is her passion and purpose, but if we want to continue the long cooperation of people and plants that feeds us, this applies to conventional farming, too. Crop rotations are key to farmers maintaining the health of their soil—the medium that sustains their livelihoods. Yet farmers don't always use them because they think they don't pay. Land, labor, seeds, and fuel are not free. Cycling through a series of crops to build soil structure and fertility, thereby limiting the pressures of weeds, pests, and disease, also costs money. Helping to establish routes for farmers to make secondary crops pay their way is what keeps Elizabeth working so hard on small grains.

For example, buckwheat is an ideal rotational crop. A distant relative of rhubarb, the plant differs enough from both grains and vegetables to be a good break from the diseases and pests these crops invite. Additionally, buckwheat improves soil tilth, as its beautiful mesh of root hairs separates and fluffs the ground. However useful, though, it is a tricky crop to harvest. The more OGRIN and its partners can help farmers figure out ways to handle and sell crops like buckwheat, the better equipped they are to serve the soil and their bottom lines.

Elizabeth has visited diversified farms in France, and tells their stories as models for American farmers. (Her trip coincided with Julie Dawson's work, and Julie took her to farms where she was working.) Elizabeth is a fan of these farms' multiple streams of value-added pursuits, from bread baking to pasta making. She likes to see farms adding animals to their system, because pasturing lands necessitates planting a variety of plants, and the animals produce fertility in return. Underseeding legumes like clover in cash grain crops (a kind of intercropping), or planting multiple crops at once, is a practice she'd like to see expanded.

These goals become realities through the joint efforts that OGRIN pursues. Her work with the Greenmarket Regional Grains Project illustrates that critical piece of farm support Elizabeth seeks to arrange.

As Greenmarket prepared to make a rule for bakers about using more locally produced flour, and was surveying available regional products, June Russell found Elizabeth Dyck, and a strong affiliation began. In grains, OGRIN and Greenmarket were natural partners. Elizabeth had contacts with growers and an understanding of farmers' needs that informed Greenmarket's quest for regional flour and grains.

As Greenmarket tried to amp up the flour supply for its bakers, it realized it had to provide support for grain farmers.

"If we could convince growers to take that risk, put them in that position, we really had to up the ante and create a bigger market," June said. They needed to draw in chefs, distilleries, and breweries so that farmers had people who wanted to buy all the grains they grew, not just bread wheat. What began as an attempt to align bakers with Greenmarket's mission took on another dimension.

The pivot took place in 2009, as Greenmarket got ready to implement the rule for flour usage to its bakers. Greenmarket and OGRIN planned for a meeting on grains at the International Culinary Institute. The event would highlight regional grains for a broad base of clients and help develop a grains system, not just a flour system.

Greenmarket has access to an influential group of potential customers. Chefs like Peter Hoffman and Andrew Tarlow began shopping at Greenmarket years before the word *locavore* codified the farm-to-table movement. With more than forty markets in the five boroughs at the time, Greenmarket was a highway for local foods. June, who has a background in the restaurant world, served as a liaison between farms and the growing number of farm-to-table restaurants whose chefs and owners were dedicated to utilizing ingredients grown in the region.

Elizabeth for her part coordinated with farmers and researchers to supply heritage and modern grains. Because the heritage varieties were available only in small quantities, she ground over 200 pounds of flour for the event on a tabletop mill. June pulled in chefs and bakers from the city who were eager to work with locally grown ingredients, as well as other potential grain buyers such as Tuthilltown Distillery, which was already making great bourbon from local organic corn.

Bakers Jim Lahey and Keith Cohen baked bread from the heritage wheat that Elizabeth milled. Chef Patti Jackson made pasta. Representatives from Sullivan Street Bakery, City Bakery, Balthazar, Blue Hill, Diner and Marlow, Hot Bread Kitchen, Gramercy Tavern, and Orwasher's Bakery attended the event. The room was filled with potential grain buyers and allies in the farm-to-table movement: Don Lewis, Glenn Roberts, Eli Rogosa, Joel Steigman, Hawthorne Valley, and restaurateur Andrew Tarlow came. Farmers were invited, too, as was Greg Mol, the miller from Farmer Ground Flour. Wheat-wise, Warthog was a standout star, and as stories of the event spread, demand for this flour buzzed. The only problem was that none was available. The entire 2009 crop of this modern variety of hard winter wheat was already spoken for.

Participants were given a list of contacts for grain growers and millers in the region. The event was noteworthy because it was one of the first occasions in North America to incorporate both a sensory description and a tasting of different varieties of wheat. This event firmly planted local grains into the conversation about local foods and served as matchmaker for growers and buyers who gave a lift to the local grains movement. The purpose of the meeting was to fill the room with potential stakeholders of a regional grainshed, those who have an interest in developing a local food system and are committed to using ingredients grown by local farmers.

Christina Grace, from the New York State Department of Agriculture & Markets, was also at the tasting. She was working on a similar initiative. New York Farm to Factory sought to help farmers access markets by arranging connections with buyers interested in sourcing locally grown ingredients. The focus was vegetables, with some interest in grains and flour. Once Christina Grace saw the momentum in grains, she proposed a grant that would take this framework and apply it to flour. New York Farm to Bakery was the result.

I attended a field day for Farm to Bakery at Wild Hive in July 2011. At the time, Don Lewis was running a bakery and café a few miles away from his mill. Upstairs in the old dairy barn, people stood on the edge of a wide, bright open room. There were farmers and millers and bakers, and the organizers who brought these people together. Sun shone through the windows and skylights onto people's shoulders.

The farmers looked like they had just gotten off their tractors. The millers and bakers were mostly men, and the organizers mostly women. Many outfits had a downstate flash, but this metropolitan dressiness was tempered for a day in the country. People wore lozenge-shaped thin-framed glasses. Beards and chunky-framed glasses weren't yet in vogue.

One person at a time spoke and everyone else paid attention. Given the earliness of the hour and the vaulted ceiling in the converted dairy barn, this felt like church. Just like a Sunday morning, people came ready to hear inspirational and guiding words. There was that same charged atmosphere of intent, all of us waiting to hear something that mattered, something that might change things. Farmers explained how they grew grains and what challenges they faced, like fusarium, a fungal infection that causes grain to develop poisonous mycotoxins. They talked about trying to get grains with high enough protein to bake bread.

After lunch at the café, bakers talked about working with local flours. They'd brought samples of muffins, brownies, and breads, and reported on how it worked, and how they might use local flour.

"About 10 or 15 percent of our customers are tuned to localization, and we can sell the higher price point of a local product to them," said Terrence Geary from Orwasher's, a bakery that's been wholesaling in New York City for a hundred years. "Some of the caterers are picking up on the interest, too. Their customers say, 'I want local products at my wedding,' and that means everything, including bread."

Orwasher's new owner, Keith Cohen, was looking to distinguish some products in the bakery's solid line by adding artisan breads. These were good for whole-grain flours, and he developed a miche and other loaves featuring flour from North Country Farms and Farmer Ground Flour.

Greenmarket's access to consumers and chefs interested in local foods is unparalleled. Shoppers come ready to buy the region's bounty, and are hungry for facts on how to use what they buy. Individual markets highlight grains or flour that vendors are using by making dishes people can sample and offering recipe cards. The agency acts as a vendor of flours and whole grains from Northeast producers, and brings in chefs and home cooks to share their passion about grains.

Special events like Grains Week, held late in 2010, introduced local grains and flour to a wide wave of New Yorkers, with tasting and cooking events at markets throughout the city. Four years later, a very different group of New Yorkers encountered grains through Greenmarket. During Craft Beer Week in February 2014, Brewer's Choice highlighted New York State grains. Valley Malt and Farmhouse Malt malted the barley, wheat, and rye that the twenty-five participating breweries used in their beers. The event was a fund-raiser for the Greenmarket Regional Grains Project.

Five hundred people populated this little city of beer on the first floor of the Wythe Hotel. The excitement for craft beer is hardly ever restrained, but this night was the peak of a weeklong beer festival, so people came primed to have a good time and ready to salute their favorite brewers. After a couple of hours of doing so, people stopped for a round of thanks. Organizers Jimmy Carbone and Kelly Taylor turned the spotlight on locally grown and malted grains, and the crowd unleashed raucous waves of appreciation. The evening was a calling card for regional grains, putting the idea to the lips of motivated consumers.

June Russell and Greenmarket leave calling cards on an everyday basis, too. When Hale and Hearty Soups, a small chain that makes scratch soups, wanted to feature a whole grain, the chef asked June what to use.

June works with Greenmarket Co., the distribution arm of the organization, to move grains through New York, too. Emmer flour was available for sale through Greenmarket Co. for months, but not much sold until June and Brian Goldblatt, the sales manager, sat down with chefs at a couple of restaurants. They brought emmer noodles to back up their claim that this was a superior flour. The tastings sold the flour and put it on the menu.

Chefs trust June for two reasons. She's known in the New York food world because she helped launch the restaurant Prune. In her current post, she's an encyclopedia of farms. She can facilitate farm-to-table connections because she is familiar with the needs of both farmers and chefs, and can do some translating to help these disparate businesses understand each other.

June has the no-nonsense veneer of someone who keeps a lot of balls in the air at once. She is sturdy and open, tough and friendly. She wears glasses and keeps her curly salt-and-pepper hair short. Her work with grains is only a fraction of her duties, but she gladly gives it a lot of her life. She's a full partner with Elizabeth Dyck on more than one grant. While inspecting farms, she also travels to grain farmers, millers, maltsters, and bakers. She documents how people are working with grains through videos and written interviews, creating resources for others who want to work in grains.

June grew up in southwest Michigan and fell into a food career by default. In college she developed a progressive consciousness. She worked for Greenpeace in the late 1980s. People around her were exploring the connection between diet and health, combined with an environmental perspective. She saw that people living with AIDS were living longer by changing what they ate; June took that into consideration as she taught herself to cook. She wanted a job at the co-op in Ann Arbor because people there understood the impact of food production on their own health, and that of the environment.

The co-op didn't hire her, so she took a job cooking at Zingerman's, a deli that has bloomed into a gourmet food empire. She started an outreach program and began buying produce for the kitchen from local farmers in 1993. Later in the 1990s, as national organic standards were developed, she brought in letters from the Center for Food Safety to her co-workers, and some of them took an interest in the issue of defining organics. For the most part, though, price and convenience were the gatekeepers for shopping choices, and chefs generally didn't think that how something was grown could affect the way it tasted.

When she moved to New York City in 1997, she continued to work in food. Cooking was what she did for a living, but she wanted to do more. She was frustrated that chefs she respected didn't see the connection between food and agriculture. Most didn't acknowledge that fresh produce, let alone organically or locally grown produce, was different from what was coming in by the caseload from conventional sources.

In 2002, she took a job with the Lower East Side Girls Club, which had a grant to start a farmers market. Initially, friends from her food life couldn't understand the shift. As the realm of good eating embraced the gospel of local food, though, her choice seemed prescient and dreamy. People now envy her job, and she loves her work.

Coming out of the food business, and having a deep respect for farming, she is well poised to facilitate change. June can see the moving parts of the food system and act as an interlocutor among farmers, distributors, restaurants, and other buyers.

Restaurant people turn to June because she knows the territory. She understands chefs' priorities. She didn't love learning the ins and outs of beverage service, for example, but she knows what bars and restaurants need from start-ups that use local malt. As magical as the word *local* may seem, local products will go nowhere if they, and their ordering and delivery procedures, don't fit into the compressed timing and cramped storage spaces of New York venues. Values-based enterprises won't float on goodwill alone. They need good logistics. That frustration she felt as a cook was preparation for this opportunity.

"I'm really grateful for that time spent in the industry because it just informs my work so much," she said. "At a certain point, I realized, okay, I was there for a reason."

Another part of her life almost eerily resonates in her work. Family members had farms, and her father, who died when she was nine, worked at Southwest Michigan Cold Storage. As she grew into her position at Greenmarket, she learned how her dad's work and the facility helped develop and support farms and orchards in that part of the state. Without refrigeration, the fruit belt of southwest Michigan couldn't serve markets in Detroit, Chicago, and St. Louis.

Seeing how her father's work fit in the larger picture of regional agriculture suggested parallels to what she could do in New York. Cold storage was necessary infrastructure for farmers to get their raw products into the world, as important as transportation and distribution. Infrastructure, she realized, was what was missing in the farm-to-table conversations around her, and in the media. In 2006 everyone focused on what farmers could do, kept talking about how farmers could change agriculture, but June saw the limits of that thinking. Farmers can only do so much. Regional farming needs the stepping-stone facilities for processing, handling, and distribution that make the national and international food systems function.

"There was this whole tier of processing facilities I wanted to advocate for because of that," she said. "If we want to think in terms of real systems being functional, you have to have these facilities."

The grains work presented an opportunity to push the conversation, and help build the food system. For grains, that means storage, mills, and malthouses. When Greenmarket instituted its local sourcing rule for bakers, it meant more than setting proportions and checking receipts to see that purchase ratios were met. June and the project actively built a network around grains, feeding all the businesses involved: farmers, mills, and value-added grain-based start-ups, like malthouses, breweries, and bakeries.

Vote with your fork is the commandment for nourishing local farms, but that sentiment misses the fact that expensive infrastructure is necessary to make change happen, and to load those forks.

June is also working on other elements that have to be in place, namely information for both producers and consumers, as was the case with grassfed beef.

"Ten years ago, people thought grassfed beef was terrible, and some of it was," June said. "Chefs didn't know how to cook with it, and farmers didn't know how to produce it. Now people talk about growing the best pastures for grazing and chefs have learned about handling. In general, people's palates have adapted to the flavors of grassfed. We know a lot more about the nutritional aspect of pasture-raised meats, and the environmental impact of raising feedlot beef."

Grassfed meats emerged as the clear winner in flavor and nutrition, but it took some time for the product and the public reception of it to mature. Grains could have the same slow progress—or the pace could accelerate as people grow aware of this unexplored territory, thanks to chefs like Blue Hill's Dan Barber, whose book *The Third Plate* is putting people's minds on farming and grains. The present-day concerns over gluten intolerance and highly processed food are drawing attention to alternatives to mainstream wheat products, too.

I asked June to guess how all of this grains work might unfold.

"Some say we're going back to the future," she said. "But our agriculture is different today. It's not the same as it was three hundred years ago when settlers came and threw down rye. We've learned something from conventional agriculture, and from thirty to forty years of organized organic agriculture, as well. In addition, we've never had this kind of food culture in the United States. This is new food territory."

The food culture and agriculture are beginning to intersect. Almost every menu in New York City has local ingredients. Some of them even name individual grains. Farmers are growing different varieties of vegetables and raising specific types of animals for chefs and other markets. Interest in breeding for flavor, instead of yields, is driving research. The potential for further connections is vast.

"We're still very much creating American food," June said. "If you look at the classic food cultures, they are, and they have had to be, absolutely rooted in their own agriculture. Because our present food culture evolved after the Green Revolution, we've had this luxury of being able to pick and choose ingredients from across the globe, but we now recognize there are problems with a global food industry."

Julia Child and *Gourmet* magazine caught and pushed waves of awareness to cuisine. And while local eating can seem trendy, June thinks it's bigger than that.

"The connections that are happening are deeper," she said. "Who knows what is going to happen in the food system. We may become more dependent on regionalism. I do think because of what we lost to globalization, people are committed to supporting the local butcher, or the local charcuterie maker, or the local beer."

So go ahead, vote with your fork. Just remember the farm advocates like June Russell, Julie Dawson, and Elizabeth Dyck, who are building the farmer know-how and the infrastructure we need for a regional food system. Thank these women for reaching backward and forward through time, linking people and plants. Thank them for taking those amber waves of grain out of an anthem and making them into a reality, into kernels that someone can make into bread.

CHAPTER FOUR

KNOW YOUR FLOUR

*S*tanding at the stove, staring at pancakes, I squint and blur my vision. This is a kid's game. Make your eyes a kaleidoscope and play with what you see.

I can picture the steps it takes to get from dirt to here. Knowing your farmer is one thing. Knowing a farmer's work is another. I think of Thor sitting in a tractor, getting a field ready for planting. He twists in an old and broken seat, looking both forward and back, checking to see where the cultivator's been, and where he's headed. Once the field is prepared, next comes the seed drill. This isn't literally a drill, but rather three long bins on top of a row of metal disks. The disks turn like circular knives, cutting furrows into the ground. Tubes run from the bin to the center of the disks, dropping the seeds at a certain rate. Heavy rubber wheels roll behind the disks to keep the depth of the planted seed constant, and to firm the seedbed. After planting, the field gets another pass with the tractor and a heavy steel roller that helps consolidate the soil further, push down stones, improve seed-to-soil contact, and keep the plants anchored through the winter.

I have helped fill the bins, and I can hear the rush of seeds, pouring like a kind of water, from the bag. I've sat in the cab as Thor planted. That bouncing up and down, the forward and back twisting—what an uncomfortable ballet. After a twelve-hour day, Thor says he feels completely beat up.

I have visited my favorite wheat when the seedlings have emerged, and the roots are taking hold before winter. Mornings, I looked at dewdrops on the thin green fingers, kneeling down and puzzling over these tiny grass plants. How ordinary, yet how magical, too. The field of soft white wheat didn't look like a lawn. The plants were a few inches tall, and each had a few blades braving the chilling weather. Rows of green were divided by dirt. Lots of rocks dotted that dirt, as if the farm were growing stones. Beyond the fields, the hills of Newfield cut like a fence between land and sky.

Small grains are planted in the fall or in the spring. Fall planting serves a few purposes, like giving crops a head start on weeds, and protecting soil from erosion. The goal is to have enough plant growth so that the seedlings can keep hold of the ground all winter, ideally resting under a blanket of insulating snow. If the roots are too short, cold nights and warming days in February or March can push the plants up from the dirt. This is called frost heaving, and it works like little earthquakes.

Weather can make other trouble for plants. One January, I saw pictures of my favorite wheat locked into sheets of ice. The fingery plants were the color of toothpicks. Though the plants had grown the perfect size before winter, early thaws stole the blanket of snow, and freezing rain made ice ponds. Thor wasn't sure if he'd lose whole fields to winterkill. He didn't plow the fields under, though, just went along as usual, frost-seeding clover into the crop. Frost heaving can uproot plants, but freeze–thaw cycles also pull small seeds like clover down into the soil. Organic farmers often plant clover over dormant grains because the practice allows the clover to germinate early and get a jump on weed seeds that are in the ground, helping the wheat shade them out.

The fields that looked so bad in January and February were doing okay in May. However, because of the challenging winter, the wheat plants were stunted. On a normal year, the wheat would grow up high enough to shade out the clover and keep it from growing too much. By July, a wet season had helped the cover overtake the wheat heads, which made harvest tough. Usually the clover hugs the ground and stays out of the way of the cutter bar on the combine. This time, Thor and his farm crew had to pressure-wash the inside of the combine because the green clover kept gumming up the works. Yields were half of what was expected, and the kernels were small, not plump. Still, the wheat made it to the bins, and later the mill, and eventually my griddle.

Wheat grows in forty of the fifty states. More land in the world is planted to wheat than any other crop. On average, 20 percent of the world's calories come from wheat; in some places wheat and bread account for a much greater portion of people's diets. This was certainly the case in the past. Fifty percent of the American wheat crop is exported. Most is grown in wheat belts in the western and Plains states. Many wheat farms are large, ranging from 2,000 to 5,000 acres. Two thousand acres is an area equal to about three square miles. Wheat and other grains grown for the commodity market often leave farms in tractor-trailer trucks at harvest, traveling to a grain elevator. Farmers take an offered price, unless they have storage facilities themselves and can wait for a better deal. Commodity prices are set by boards of trade and reflect international markets. The prices can swing wildly and have little to do with what it costs a farmer to grow a crop.

There are six classes of wheat: hard red winter, hard red spring, soft red winter, durum, hard white, and soft white. *Red* and *white* refer to the colors of the bran. Red wheats have more tannins than white wheats, and these tannins tend to have a bitter flavor. Hard wheats have more gluten than soft wheats, so hard wheats are preferred for making bread. Durum flour is used for noodle making, as are white wheats, much of which are exported for the Asian noodle market. Interest in healthier, whole-grain eating has driven a market for white whole wheat flours in the United States, too.

Wheat kernels, like most grains, have three major parts: the outer layers of bran, the endosperm, and the germ. Wheat kernels are shaped like teardrops, and the germ sits at the base of the endosperm. As the edible seeds of certain plants in the grass family (Poaceae), the first job of wheat is reproduction. The germ sits protected, waiting for moisture and temperature cues to penetrate the protective layers of bran and signal the beginning of germination. As the seed sprouts, the endosperm is food for the growing plant. We humans want that food, too.

Bread wheat is hard wheat, meaning the endosperm, the starchy part of the grain, is very hard. Pastry wheat is soft wheat. Hard wheat has more gluten than soft. Gluten is made up of proteins, gliadin and glutenin. When these proteins are mixed with water, they create strong bonds—a matrix that is a good skeleton for bread. Over the last decade, American interest in gluten-free diets has soared, morphing the low-carb trend that began with the Atkins Diet in the late 1980s into the current fear of gluten. Before anyone shuts the door on wheat, however, I think they should get to know what frightens them. Here is the monster that I adore.

The habits of wheat have compelled us for more than ten thousand years. The long shift from hunting and gathering to settled life began in the Fertile Crescent in the Middle East. Our ancestors began to plant some of the seeds they collected, and beans and grains became our foundational crops. Wild barley and einkorn were among the first grains domesticated. Whether the plants domesticated us, asking us to settle down and tend them, or we domesticated them, asking them to sit still and feed us, is up for debate. Other questions about grains are more hotly debated. Should we eat grains at all, or follow Paleolithic prescriptions and seek foods that are the closest things to woolly mammoths and wild tubers? Won't ancient grains free our bloated bellies from the burdens of gluten? I am prone to romancing the past, but the past might not cure us.

The food system is complicated and our lives are crowded with work and information. What facts we get arrive in bite-sized nuggets or polemicized epics, so it is hard to make sense of farming and nutrition. Corporate battles for seed ownership, which are tied up with GMOs, add to the confusion, making any involvement of science in agriculture seem suspect. There is no GMO wheat on the market yet, but genetically modified corn, soy, and other plants sit unidentified in many foods. Monsanto and other corporations aggressively fight consumer battles to label GMOs. Farmers have been sued because their corn showed evidence of patented genes the wind had blown onto their fields. When agribusiness bullies farmers, science, by proxy, gets some of the blame.

One of the popular critiques of wheat is that it has been bred for high yields, and bred to work better in industrial baking. Modern wheats are accused of causing sensitivities to the grain, and some believe they are responsible for rising rates of celiac disease. Ancient and heritage wheats, it is proposed, might not have the same negative side effects.

Do older wheats offer more food value? Are heritage grains our salvation? We eaters need a lot of help figuring out what is good and what is bad. Scientists, agronomists, and nutritionists are trying to determine some answers.

In my ongoing quest to understand the complicated territory of grains, I turn to a lot of people. Many of them are at Cornell University. Faculty and researchers are ready resources. A great way to hear their thoughts is at field days, events that bring together farmers, extension agents, and food advocates to learn about research projects.

Every July I go to an organic grains field day in Freeville, a research farm north and east of the university. Little distinguishes this farm from the ones around it, except extra signage, and the fact that fields are patchworked, divided into ribbons of test plots.

This year, late on a hot, bright day that was supposed to break into storms, I toured the organic grain trials with about twenty other people. Forty different kinds of grains, tall and shorter, green and bluish, spread out before us, a quilt of possibilities. These ancient, heritage, and modern wheats, spelts, and barleys were close to ripe. Most of the stalks were still green, but the heads looked yellowish and tan. The grains were beginning to dry. In a couple of weeks, everything would be harvested.

"This was a cold winter," said David Benscher, who researches organic grains. "We planted in late September and early October. December was 4 degrees below normal. January was 5½ degrees below normal, February was 7 degrees below, and March was 8 degrees below. In April I stopped looking."

David stood in front of a nearly bare plot labeled RED FIFE. A few stalks stood up brave and alone. Neighboring plots were thick with plants.

Red Fife wheat was revived in western Canada after languishing in seed banks for nearly a century. This was the dominant bread grain for a good chunk of the nineteenth century in Canada. Growing out the variety was fairly simple. The plant did well for farmers, and because of its relatively high protein content, the flour fit well into current markets. Bakers and consumers thunder for this type of wheat, giving hope for the restoration of older grains. While farmers in Ontario have also had success with Red Fife, this particular planting had not fared well. Two grain heads bent at the neck, as if ashamed by their performance.

"We experienced some winterkill," David deadpanned, sweeping his hands at the dismal plot. "But most varieties came out of the winter pretty nicely."

Our group of farmers, bakers, and researchers laughed. Elizabeth Dyck noted that Red Fife is facultative, meaning it can be planted as a winter or spring grain. Traditionally it is planted as a hard red spring wheat, but

Elizabeth said that Red Fife had been successfully grown as a fall-planted crop in the more southern regions of the Northeast.

The questions about heritage varieties started to volley. The purpose of heritage grain trials like this one is to identify seeds that will grow well in the Northeast, and to find the next "it" grain for a consumer market that's revved for something old-fashioned and, presumably, better for health.

Red Fife, and White Sonora wheat in the Southwest, have undergone swift reintroductions, but these two heritage varieties are the exception, not the rule. Farmers left behind old seeds as plant breeders crossed types with desired traits. Science did not just sweep in and discard perfectly good plants, like worn-out clothes; new fashions of seed were sewn from old. Removing old grains from seed banks can be like trying to wear your grandmother's bridal veil, a romantic but futile effort.

Red Fife's success is related to its protein levels: The variety sometimes clocks in at a stunning 13 percent protein. Hitting high protein in the Northeast is difficult for any variety, regardless of the marketing merits assigned to age. White Sonora is stunning farmers, millers, and bakers partly because the Southwest is arid, and the plant is thriving in that climate.

Heritage varieties can be defined as those developed and grown before 1950; grains bred afterward are considered modern varieties and generally contain dwarfing genes to produce shorter stalks. This trait was selected to make the plants receptive to high fertilization, to boost protein levels. Heritage and ancient grains are tall and can't take a lot of fertilizer, because the stalks would "lodge," or fall down in the field. These and other plant trials seek to identify varieties that do well in northeastern climates and soils, and can use the advantage of height to compete against weeds.

Science is steady and slow. The race of questions at the field day made me wonder how plant breeding programs can keep pace with demands, let alone predict the next wishes people have for food. The researchers answered questions as best they could, but graduate student Lisa Kissing Kucek urged people to hold their thoughts about gluten. She would be giving a presentation later on gluten and grain varieties. The skies were filling with clouds, and she wanted us to see the test plots while we could.

Gary Bergstrom from the Department of Plant Pathology spoke about fusarium head blight, the fungal infection that creates vomitoxin in cereal grains. Fusarium can be a dealbreaker in the grain industry. The USDA sets tolerances for DON, also known as vomitoxin, at 1 ppm for milling and malting grains. This is one of the biggest challenges in the humid Northeast. The virus travels on the wind, and also overwinters in crop debris from corn and other grain plants.

"Rotations and other cultural practices, and variety selection, can offer some control," he said. Fungicides applied at the onset of flowering can help in conventional systems, but there is no option for organic crops. Grain cleaning can help remove the small, scabby kernels and reduce the toxin levels.

Gray clouds roiled, and the researchers hustled us through the plots, pausing by some tall and bluish spelt plants that waved in the building tumult. This was a sea of grain.

We crowded under a tent. There was debate over whether to move to Cornell. Rain began falling, straight at first, and then slanted. We gathered closer together, pulling tables of pamphlets and bread away from the water, trying to keep them dry. Lisa Kissing Kucek started her talk.

"My colleagues and I have spent a happy year evaluating the literature on ancient and heritage grains," she said, speaking loudly so as to be heard over the pounding rain. "Consumers and farmers are always asking us for information on the perception of health benefits. Glutens are storage proteins that seeds release upon germination and growth. Normally, wheat is very nutritious, but people can have problems with the proteins in wheat. Celiac disease is one of the most common autoimmune responses to gluten."

Lightning cracked. Lisa stopped talking. The rain turned to hail. The hail got bigger, and we all stood still, nervous and suddenly cold. Over the raucous din of the weather, someone said we were foolish to be near tent poles. We decided to head over to campus, where we could see dehulling equipment and sample some beers made with local grains.

Following each other's taillights, we drove through sheets of water. By the time we got to a series of long buildings, the weather had worn itself out. The sky was clear again, but the parking lot had puddles. The buildings were labs and storage garages, a very functional, ivy-free part of academia. We were soaked and quiet as we gathered in one of the garages. Now that we were indoors, we could joke about the storm and its condemnation of the topic of gluten. People cracked beers and laughed. The beer was supposed to be for the end of the event, but the trauma of the hail entitled early fortification.

I stood near wooden crates labeled CEREAL in big block letters, a lettering style I associate with the army, thanks to the TV show *M*A*S*H*. The crates were filled with seed samples. The setting felt more rustic than scientific, but I know that's a symptom of my being a guest in the realm. I may never quite catch up to what is normal for farmers and scientists.

Brian Baker, an agricultural economist, continued the program and the thread of debunking idolized grains. His topic was dehulling, the removal of hulls from grain kernels. Spelt, emmer, and einkorn all require dehulling before we can eat them. The Romans, Brian said, gave up on ancient grains because of the extra work involved. He demonstrated a couple of dehullers, including a prototype built by a class at Cornell. Brian asked growers to think hard before investing in the infrastructure and labor needed to process ancient grains.

I had to leave before Lisa gave her presentation. I was too chilly and drenched, and knew my questions about wheat couldn't be answered in what little time remained. I planned to contact Lisa later for a one-on-one talk.

My doubts about wheat began in the 1990s when I lived in Seattle. Half the people I knew were eliminating wheat and or dairy, following wisdom that came from Bastyr, a naturopathic medical school. Eventually I began my own wheat-free days, trying to get rid of eczema that had spread all over my body. I itched so badly that I dreamed I needed new hands and arms. I avoided wheat for a couple of years, and remained afraid of it for many more, baking with a combination of soy and rice flours to keep pancakes in my life. I also avoided sugar, alcohol, coffee, and nuts.

I was never sure what helped, but my eczema went away. The social fear of wheat spread beyond the Pacific Northwest and became a mainstream phenomenon. As more and more people gave up gluten, I fell deeper in love with flour, having discovered locally grown, stone-ground grains. Somewhere between my conviction that stone-ground, whole-grain flours of known origin will save us, and the general population's disdain for a foundational food, there has to be a logical approach to gluten. I made an appointment to meet with Lisa Kissing Kucek and hear what she'd discovered.

We met in a lab at Cornell, sitting on stools near black counters that held microscopes and other analytical tools. A bookshelf held grain textbooks and a fat binder that contained a single day's papers presented at a conference on soft wheats. An entire binder on a single topic? I felt humbled by how much there is to know about grains.

"The world didn't want to know the truth about gluten," Lisa joked about the field day. Lisa is in her late twenties and working on her master's at Cornell. She wears functional clothes that tip toward the side of fun, like an orange T-shirt with pleats. Her brain is like that, too, practical and warmly curious.

The heritage wheat trials at Freeville had inspired her to study the scientific writings on gluten and wheat varieties. Vintage wheats are captivating the imagination of the public, but is this a case of the emperor's *old* clothes? Lisa wanted to see what is actually known about the nutrition and safety of different wheat varieties. She and her colleagues looked at over two hundred research papers, undertaking a literature review of research published on the topic, *The Grounded Guide to Gluten*.

"I wanted to see what wheat actually does in the human body," said Lisa. She began with gluten, but felt that focus missed the mark and stretched her lens back to include the broader subject of wheat. "I wanted to see the differences in reactivity of wheat, among ancient, heritage, and modern wheats. And then I wanted to see the impact of wheat processing methods."

By *reactivity*, she meant how people's bodies react to wheat and gluten. Before she got into what she found, I asked her to explain some of the misconceptions about wheat. Many of them come from the book *Wheat Belly*, written by a cardiologist, William Davis.

"He calls modern wheat a chronic poison and claims that Norman Borlaug [the father of the Green Revolution] hybridized wheat," Lisa said with some impatience.

Her frustration was familiar. I'd heard nutritionists and cereal scientists defend wheat from the book's indictments, too. However, I didn't really understand why the claims made were wrong. What, I asked her, is the problem with hybrids? To explain, she gave me some basics about reproduction in grains.

Wheat is a selfing, or self-pollinating, species, meaning it reproduces its own genotype over and over again. Corn is not self-pollinating, and neither is rye. If these species pollinated themselves, the resulting plants would be inbred and weak.

Hybrids are the plant babies created when two distinct plant parents are crossed. When a wheat mother with the genetic combination (or alleles) HH for tall height is crossed with a wheat father with alleles hh for short height, the first generation of plant offspring would be hybrids with one set of genes from each parent, Hh. Hybrid corn is sold in this way, as the first generation of a cross between two different corn parents. Wheat, however, is not marketed in hybrid form. Wheat breeders like Norman Borlaug allowed the first-generation crosses to reproduce themselves for at least six generations to get a stable plant, fixing certain desired traits from the initial plant cross. The highly vilified modern wheats are plant crosses. Norman Borlaug took a dwarfing gene that naturally occurred in wheat and made crosses to produce shorter-stalked plants. From those crosses, Borlaug let wheat grow out over multiple generations, and then selected among the thousands of great-great-great-great-grandchildren to make a line of wheat, a variety, to release to farmers.

If you save the seeds from a field of wheat, what grows in following years will have a great deal in common, genetically speaking, with the original variety. However, since corn is not a self-reproducing species, saving the seeds of hybrid corn doesn't work. What grows will differ wildly from the previous season's plants.

Looking at the commercial development of corn helps me understand why we are so befuddled about grains, hybrids, and GMOs. Corn hybrids were developed in the 1930s. GMO corn contains traits from other species, and was developed in the 1990s. Genetic modification, aka genetic engineering, uses genetic material from another species, something that would be impossible in nature.

GMO wheat is not on the market yet, but it is in development. The modern wheats developed after 1950 are bred using classical breeding techniques established in the late nineteenth century and based on the works of Gregor Mendel. Classical breeding means merging plant material, usually pollen, from one plant parent, or progenitor, with another. The simplest form of this can happen in your garden. When squash blossoms are sticky with pollen, take some on your fingertip and swipe it on the stamen (female organs) of another plant's flowers. The resulting fruit contains seeds that, when saved and planted, will exhibit traits of both plants. In labs, the anthers (pollen-producing "male" organs) of self-pollinating plants are removed. Then the pollen is transferred using simple tools, like tweezers and Q-tips, to fertilize another plant, and a plastic dialysis sleeve is placed over the fertilized plant. Using this method, plant breeders have cultivated new varieties, or cultivars (short for "cultivated varieties").

The nomenclature is confusing. Breeding gets lumped in with genetic modification or engineering, and *hybrid* is a dirty word to some people these days. There is a prejudice against directed hybridization, where people choose the plant parents rather than letting the wind, insects, or other pollinizing agents do the selection randomly in nature. This preference has come about partially because modern plant breeding programs, especially in grains, target industrial agriculture. Another strike against hybrids is that they don't reproduce themselves "true to type" from seed, making farmers reliant on purchasing seeds every year, and netting companies a great deal of control over the food supply. What this means for food, and for our ability to feed ourselves, is an enormous problem, but food sovereignty is not what I'm considering here. I want to unlace the critique of wheat as a poisonous hybrid unleashed by plant breeders after 1950.

To review, modern wheat varieties include the genetic results of plant crosses—but so do the heralded heritage varieties of wheat, which are also the results of directed plant breeding efforts that began in the last quarter of the nineteenth century. Ancient wheats, while safe from the presumed harm of human-directed plant breeding, are not the pure, unhybridized foods some might want them to be. Wheat hybridization occurred without laboratory intervention. Eight thousand years ago, the number of chromosomes in einkorn, the earliest type of wheat plants, increased through a freak of nature.

"Einkorn is a diploid like us, and has two sets of chromosomes," Lisa explained. Polyploids have more than two sets. A plant relative of einkorn hybridized with another early grain species, and the progeny had four sets of chromosomes. This next step for wheat meant bigger seed size, which would have been obvious to eaters who were always on the hunt for more food value. The ancient grain emmer, a tetraploid, was the result. "Farmers were involved. They saw this, and they probably said, *Hey, look at this giant seed, let's keep this moving.* So then you have emmer being developed, and the freak incident happened again, and then you get wheat."

Wheat. Grain kernels that fall free from their hulls when harvested and threshed. A genetic triploid with three sets of chromosomes. Here's another place where the nomenclature gets tricky, because this species is also known as common wheat. Not the wheat that's been recently bred for higher proteins to serve the needs of the baking industry, but the wheats that first had the kind of beautiful, elastic gluten necessary for leavened bread.

"Einkorn has gluten, emmer has gluten, and modern wheat has gluten," Lisa said. But only common wheat, the one that is thousands of years old, not sixty, has the right ratio of high-molecular-weight glutenin, which is the biggest factor in making bread dough rise.

Glutenin and gliadin are two of the proteins found in wheat. Each type of wheat has different ratios of gliadin and glutenin. Different growing conditions can create different ratios in the same varieties. In general, though, soft wheats have more gliadins, which have more of the flowing properties that are important for things like biscuits and pastries. Leavened breads also use gliadins; these proteins fit into the larger gluten complex that's created. Sulfur compounds in glutenins create the large branching structures inside bread dough. That's how glutens work in baking, but why are they a problem in people?

"Proteins, both plant proteins and animal proteins, can be really difficult for humans to digest," said Lisa. "We can perceive them as pathogens, something that's going to attack us. The major food allergens are all proteins."

Yet not everyone has problems with wheat proteins. How does gluten relate to celiac disease, wheat allergies, and non-celiac wheat intolerances? Lisa started with celiac disease, because it is the best understood and most dangerous of all the wheat diseases that exist.

"Anywhere from 0.5 to 2 percent of human populations has celiac disease in areas that have consumed wheat for a long time," Lisa said. "A specific genetic signature can predispose individuals for celiac disease. Certain protein compounds from wheat can become really problematic for celiac individuals."

Some of those long, difficult-to-break-down proteins can provoke a toxic response in the intestine, and start a negative feedback loop that increases the permeability of the gut wall, allowing more gluten proteins to penetrate the intestinal lining. This causes pain and discomfort, but also leads to long-term health problems. People with celiac are also more predisposed to other autoimmune disorders, and should not eat anything with gluten. Barley and rye are out, because they have gluten-like storage proteins, too. In barley these are called hordeins; in rye, secalins. Even oats, though thought to be safe if they are processed in a gluten-free facility, may be a problem; the storage proteins of oats, avenins, are aggravating to some people with celiac disease.

The second category of wheat-related disease is allergies. These are allergies to wheat proteins, which can mean gluten and/or other proteins, like alpha amylase inhibitors. Baker's asthma affects people through the respiratory system. A wheat allergy can also present itself as a typical food allergy, with symptoms ranging widely, from feeling uncomfortable to death. Another type of wheat allergy is exercise-induced. Anaphylactic shock can result from eating wheat within thirty minutes of vigorous exercise.

The third category of wheat problems is less well defined. This is the murky world where people don't have celiac disease, don't have a wheat allergy, but seem to respond well to eliminating wheat from their diets. People can be diagnosed with non-celiac wheat sensitivity, an umbrella of symptoms whose causes are not understood. Irritable bowel syndrome (IBS) is another umbrella-like title for a broad array of symptoms, some of which overlap with the symptoms of non-celiac wheat sensitivity. A staggering 14 percent of Americans have IBS.

FODMAPS, or fermentable, oligosaccharides, disaccharides, monosaccharides, and polyols, are now being considered a culprit in these disorders. FODMAPS are carbohydrates that fall into the different categories listed in the acronym. Fructans are some of those fermentables, and wheat is the largest source of fructans. In addition to fructans, the FODMAPS family also includes sorbitol, which is found in stone fruits; raffinose, found in legumes, lentils, cabbage, and brussels sprouts; and lactose, found in dairy products. FODMAPS, including fructans, make just about everyone gassy as microbes break them down in our bodies. People with fructose intolerance and fructose malabsorption experience more extreme and painful bloating, caused by fermentation releasing hydrogen gas in the intestines.

"If that's what's bothering you, and you just cut wheat out of your diet, it's going to help somewhat. But other things could be causing your symptoms, too," Lisa said.

As she and her colleagues reviewed the literature, two elements jumped out as potential culprits in wheat problems, wheat processing and vital wheat gluten.

Processing is especially suspect, from milling styles to industrial baking methods. Most flour sold today is white flour. Shedding the bran means discarding the very enzymes that are essential to breaking down proteins. Irradiation has been observed to increase the allergenicity of wheat. The Chorleywood Process, invented in 1961, could be responsible for the rise in reactions to wheat products, too. This process abbreviated mixing times through the use of additives and dough conditioners. Bleaching flour might be another problem.

The presence of vital wheat gluten in common foods is another concern. In the 1940s, this by-product surfaced as people were trying to isolate starch from grain. Inadvertently, the cheapest protein, cheaper than soy, cheaper than whey, was invented. The food industry has gobbled up this ingredient because it lends so much to processed foods, from protein and a pleasant mouthfeel to the more abstract quality of meatiness, something that makes products very satisfying to the tongue.

"Vital wheat gluten is everywhere," Lisa said. Everywhere from salad dressings to instant soups. "Look at multigrain bread, which is hot right now. Incorporating all these other grains that don't have gluten, you need extra gluten to make the bread bake."

In fact, vital wheat gluten is found in 30 percent of supermarket foods. Wheat may not be as indictable as the quickly growing gluten-free industry would like consumers to believe. How we handle wheat might be to blame for some of the dietary discomforts that so many trace to gluten. Yet we all like a good dietary straw man to set up as a target. Growing up in the salt- and fat-fearing era of the 1980s, I had to train myself to use salt, and learn to not feel guilty for loving butter. Gluten has become a handy scapegoat for what we think ails us. Besides, if less processing of food is the solution, how can the food industry manufacture us a solution? It makes more sense to point a finger at gluten, and direct critiques at plant breeding.

"Farmers were our breeders for thousands of years. They witnessed the crossing of einkorn and emmer and this crazy gigantic thing that was modern wheat," Lisa said.

Like Julie Dawson and Elizabeth Dyck, Lisa has great respect for farmers. She applied to Cornell because she'd worked in Cuba on participatory breeding projects. Looking for her next academic step in the United States, there were few options. It is disheartening to know how removed farmers are from the research process. As sustainable agriculture becomes more mainstream, this is changing. The demand for specialized grain production is creating a hunt for varieties suited to locales outside the wheat belts, and this quest is ideal for projects that incorporate farmers. Lisa is involved with one, working with farmers to breed and select grains on their farms in the Northeast.

Before farming was segregated from everyday life, breeding and plant selection used to be more familiar to people. Take a look at Iowa in the early 1900s. After reports of poor corn germination flooded the state agricultural college, Perry G. Holden led a campaign to improve corn production. Everyone involved, from politicians to seed companies, farmers to schoolchildren, shared a common vision of successful farming.

The cooperation was easy to arrange. Prior to working at the ag school, Holden had worked for a hybrid seed company. Holden and educators traveled the state on the Seed Corn Gospel Trains in 1903 and 1904 to teach people about seed selection. The lessons were so popular that people stood outside the railcars, listening. Holden went to schools, too, to teach kids how to boost corn production.

What an example of science working with the community!

Over the next hundred years, corn breeding shifted almost entirely into industry's hands, and farmer involvement gradually vanished. In the 1970s there were 250 seed companies breeding and selling corn; now we only have 5. GMO breeding could never be tackled by kids in their backyards. The Mexican campaign Sin Maiz, No Hay Pais (No Corn, No Country) has fought the creep of cheap American corn across the southern border since NAFTA: Descendants of the very people who originally developed corn are losing the ability to profitably grow their heritage varieties. We are in the thick of change.

As our relationships with plants become few and far between, our misunderstandings about them increase. We want to impale gluten, the magical elastic substance that allowed us to invent bread. Bread allowed us to push our energies beyond a constant search for food, and into building civilization. In a functional sense, gluten holds bread together. In a metaphoric sense, gluten holds us together, inviting us to a series of rituals that still resonate, even as we've grown removed from their practice.

Yes, wheat breeding has focused on increasing protein levels, largely to accommodate the baking industry. Industrial baking has changed dramatically, just like all facets of food production. Identifying protein and gluten as foes is not going to solve anything.

This is how I understand the war on gluten. After World War II, agronomist Norman Borlaug was working in Mexico, trying to develop staple crops to feed a growing global population. The genes for shorter-stalked wheats had shown up in Asia, and Borlaug used them in his breeding experiments to help plants put less energy into growing stalks; this would facilitate the plants putting more energy straight into the kernels. Shorter stalks have another advantage, too, because longer stalks tend to lodge, or get blown down in wind and rain.

Borlaug's improvement, one of the best-known parts of the Green Revolution, also allowed for the application of heavy fertilizers, which taller stalks cannot tolerate without lodging.

In retrospect the Green Revolution was a disaster, introducing high-input, high-cost agricultural methods to developing countries and stripping farmers of the ability to feed themselves and their neighbors, rather than equipping them to do so. However, that doesn't mean that all short-stalk wheats are bad.

We need not adore the old, and yet that is our human tendency, especially when the new stuff seems to be the problem. Some of the research Lisa looked at found a linkage between high-molecular-weight glutenins and the epitopes that are reactive to celiac disease. (An epitope is the specific part of the protein that's going to become reactive in the intestine for people with celiac disease.) However, some varieties of ancient and heritage wheats also had high amounts of these epitopes.

The screening has been done on European genotypes, where concentrations of celiac disease are higher. Finland has the highest rate in the world; 2 percent of their population has celiac disease.

"Some places in Northern Europe have talked about testing their lines" for the epitopes, Lisa said. Then this issue might be addressed in breeding programs, and wheats that have a high amount of epitopes would not be released for commercial production.

This seems a better way to proceed, but culturally, I think we are too infatuated with fear in America to take a slow, reasoned look at what foods hurt us. Our social concepts of nutrition are fueled more by market concerns than public health issues. The rise of the gluten-free market illustrates this swift response to dietary fads and fears. Even stalwart whole-grain companies like Bob's Red Mill, a national brand that uses stone mills and has been around since the 1970s, sells in this climate. Nearly half their sales are from gluten-free products, including flours and mixes.

Yes, diagnoses of celiac disease are on the rise, and yes, people are finding relief from a range of symptoms by avoiding gluten. But I don't think that turning our backs on a foundational crop, or putting ancient grains on a pedestal, are good long-term solutions.

CHAPTER FIVE

BREAD BUILDS
COMMUNITY

Every July, people come to central Maine to learn about breads and grain. The Kneading Conference takes over the Skowhegan Fairgrounds and makes a little world where everybody speaks some tongue of bread, fire, or flour.

The fair is old but the grandstand is new, and looks like the long mouth of an aluminum giant. This is the first thing you see when you park, and there's no hint that, behind the grandstand, the stage is set for bread. Rolling racks of sheet trays prop up long-handled peels made of steel and wood. Stacks of bowls wait on tables beside plastic tubs for proofing dough. Piles of bricks are ready for a masonry oven workshop. Bags of sand and mortar will become the mud hump of an earthen oven.

Mornings are misty and chimneys puff smoke from ovens that look like mushroom caps, lending a hobbity feel. People eat meals together at long picnic tables, granola made from local oats, and loaves from the day's classes.

For two days, people make tough choices about compelling workshops: baking with rye, or using sprouted flours? The story of grain projects in the Northeast, or Arizona? Baking teacher all-stars sit in on one another's classes like jazz musicians at a club, weighing in on questions of fermentation and timing. Serious home bakers come to perfect their hearth-style artisan loaves or laminated doughs. Professional bakers come to learn, too.

On the third day, bakers make their slow good-byes and the Artisan Bread Fair draws big crowds. People eat pizza from a wood-fired oven and take home interesting loaves. Vendors sell Maine kelp and hand-loomed scarves. These booths have little to do with the conference's topic, but relate to its gut message of reviving local economies.

Snug in the midst of lakes and woods, Skowhegan is home to eight thousand people. The former mill town wraps around the banks of the Kennebec River and its dammed falls. From the 1830s to the 1950s, people made things here and jobs were plentiful. Now the paper plant and the hospital are the prime employers, and 51 percent of the county qualifies for federal food assistance.

Author Richard Russo knows this kind of place, and so do I. He grew up not far from me in Gloversville, New York. Early in the twentieth century, my city of Troy made 90 percent of the world's collars and cuffs, and Gloversville made most of the gloves. Instead of using his hometown's particulars to explore blue-collar life, he used Skowhegan and neighboring towns to model the worn-out mill town that starred in his book *Empire Falls*. The 2001 novel won the Pulitzer Prize, and the movie was filmed in and around Skowhegan. Exemplifying the wasteland of the postindustrial Northeast wasn't the kind of notoriety everyone wanted. Some people decided to rewrite the narrative of Skowhegan, not on the big screen, but on the more lasting screen of everyday life.

"We needed something new to be known for," said Amber Lambke, who directs the Maine Grain Alliance and a slew of projects to revitalize grain production. Initially the community's revisioning efforts were aimed at beefing up the economy in general. In 2005, the town began participating in the Main Street Program, a national strategy to help struggling downtowns bolster themselves. They started a farmers market. Amber was volunteering on these initiatives when she and her neighbors came together to discuss other possible steps.

These neighbors included a fair number of bakers and masons. Albie Barden had an idea. Albie builds ovens and is the Johnny Appleseed of masonry heaters, a style of radiant heating that relies on collecting warmth from intense fires in bulky masonry structures. He'd recently spoken at Camp Bread, the Bread Bakers Guild of America's immersion workshop, and at another bread gathering hosted by Alan Scott, whose ovens and oven plans helped launch the artisan bread movement in the 1990s. Alan Scott suggested Albie organize an oven and bread conference on the East Coast.

Decades before, Albie helped start the Common Ground Fair. This is MOFGA's (the Maine Organic Farmers and Gardeners Association) harvest festival and draws fifty thousand people each September to celebrate handmade living and share the necessary tools and skills. The Kneading Conference could be a similar celebration and skill-share event based on bread and grains.

The topic seemed a perfect magnet. The artisan bread movement had sparked a need for baking instruction and a demand for wood-fired bread and pizza ovens. Bakeries from as far away as Boston were hunting for local flour. Over the course of six quick months, organizers mapped out a two-day plan, inviting experts to speak about their work in the field and teach baking classes. Don Lewis, whose Wild Hive Bakery had built a regional grain and mill project in the Hudson Valley, would be the keynote speaker.

In the summer of 2007, the first Kneading Conference drew seventy-five people to a church parking lot. Three years later, the event moved to the fairgrounds, which proved a good spot to stitch bread back to the land.

Skowhegan is the seat of Somerset County, and its fair is the oldest running agricultural fair in the country, starting in 1818. Somerset County produced 239,000 bushels of wheat in 1830. During the Civil War, Maine was one of the breadbaskets for the Union army. By the early 1900s, though, the state had lost most of its grain farming.

Amber and others who were courting a new economy saw real potential in grains.

The idea of the conference wasn't to make something happen for a weekend, but to create lasting changes. Skowhegan already had a farmers market, so people knew what kind of engine you could build for regional farming. Could bread and grains renovate the community as well?

This was my question about Skowhegan, too. If wheat was indeed the great civilizer, and growing grains made us stand still, and let us create cultures and cities, could wheat civilize us again? Skowhegan and The Kneading Conference seemed a perfect petri dish to study the proposition. Could a return to the basics of bread, and a recuperation of business relationships outside the commodity system, change a region? I think the answer is yes.

The conference draws in a huge amount of community support. Members of the board are active year-round, finding ways to fit the mission of the Maine Grain Alliance, to preserve and promote grain traditions, into everyday life. Each July, a herd of volunteers makes the workshops and Artisan Bread Fair happen. Many of these people are locals who give their heart and time to the event, thrilled to be a part of the hubbub and buzz. A good number of people come to the conference on work-study scholarships, and they earn their way by helping with logistics, moving chairs, food, plates, dough, and wood.

The workshops are beehives. People pat mud into an oven with the sweet guidance of Stu Silverstein. Pat Manley shows how to make a brick oven. There are huddled dives into details on the chemistry of sourdough and the nuances of growing heritage grains. A heady exchange of bread facts takes place, and a kind of family forms. The energy is real, and nourishing.

The momentum of these connections sticks around Skowhegan even after most conference attendees return home. Early on, organizers realized that to make a real impact on grains and bread, the town needed a mill. Without a processing facility, local flour would remain just a theory.

Amber Lambke and baker Michael Scholz decided to take on the project. Initially, they visited established mills, hoping to find someone who wanted to start a second outlet. Eventually, they accepted what was becoming apparent. If they wanted a mill, they'd have to start one.

"We went to every mill in the Northeast," Amber told a group of people touring the Somerset Grist Mill after the conference in 2014. The story of how the mill happened was as important to her as describing how the equipment worked. "We saw Jack Lazor and visited historic mills in Rhode Island. We went to New Brunswick and Aroostook County. We heard over and over again from sixty-year-old men, 'If I die tomorrow no one would know what to do.' So from the get-go that was our goal, to make something that would outlast us."

At home, they scouted locations. The Somerset County Jail was for sale. The site was not an obvious fit, but after Amber took a grain milling course at Kansas State University, she realized the jail would work well. Because milling relies on gravity to help move grain, they could use all three stories of the structure. Plus, the jail's walls were 3 feet thick, which would insulate downtown from the vibrations and noise of milling.

Today the brick building doesn't seem like it was ever a jail. Three tall slim silos stand near the street, looking a little like galvanized steel pencils.

A bulky yellow contraption frames the garage door of the loading dock. This is a cyclone, a stationary vacuum that controls dust from milling.

About thirty of us gathered in the parking lot outside the mill for a tour, which began on the ground level, in the "in and out room." Inside, a forklift sat at the ready, and bags of milled flour and rolled oats filled racks. I stood near a machine called a gravity table, which is used to sort out seeds and grains of different shapes and sizes.

The town council, Amber said, had initially rejected the proposal for a mill, hoping that something more glamorous would come along, like a restaurant or a bar. Finally, the plan for the mill was the last one standing, and Amber and Michael bought the building in September 2009. They took time to gather equipment and resources. Their goal was to enrich the community, not the enterprise, so avoiding debt was even more important than for other start-ups. Amber pursued funding with creativity, identifying the broad benefits that the mill could bring. The project received a Community Development Block Grant because of the potential for job creation. Slow Money Maine, part of a national organization that supports good food businesses, was a natural ally.

"The idea of the mill came along just as this investment scheme was developing," said Amber. "Since this is an infrastructure business that supports farmers on the upstream and other businesses on the downstream, we were taken under the wing of the strong network of people who wanted to make mission-based loans."

Alternative investment routes like Slow Money got a boost in the wake of the Bernard Madoff fiasco and the bailout of Wall Street. Other support came from grassroots fund-raising, like a Kickstarter campaign that gave the final nudge to get the mill open in September 2012.

We walked single file up a tight stairwell to the mill room. The mill is an Osttiroler, which looks like a beautiful piece of furniture. These Austrian mills are made of a composite stone and set horizontally in a white-pine-planked frame.

I stood in a side room, next to the old wooden seed cleaner, which was lovely in another way. The Clipper cleaner has that old-timey appeal of farming antiques, but the technology is up-to-date. The mechanics for air-screen cleaners haven't changed for a hundred years; newer models just have shiny metal casing. Both kinds are rectangular cubes that hold sets of screens to sort out field debris and weeds.

I looked at the walls of the old jail cell, which were painted yellow, unlike the first time I visited. Then the walls were gray, and scratched with graffiti, people bad-mouthing one another, declaring love, or just writing their names. When I saw those reminders of the lives that had been held in limbo, I felt guilty for being excited about the building's new use and purpose.

Now the tasks of the milling equipment set the tone. This is a place to make flour. I wasn't as giddy as I was that first time, but fresh flour is still novel enough to make people squeal, like the miller, Michael Scholz.

As people in our group found places to stand, Michael realized he was standing next to Peter Reinhart, who had taught a class at the conference.

"Making bread for the first time in 1996 was a really big deal for me," Michael told us. "I used Peter Reinhart's book *The Bread Baker's Apprentice.* So I'm very, very excited that he's here."

The miller took a moment to compose himself, running his hand over his short salt-and-pepper hair. He smiled up to his eyes. Even his wire-framed glasses twinkled as he started to explain the mill.

"This came from Austria and has two 4-foot granite stones," Michael said. "The top is the runner and the bottom is the bedstone. The hopper feeds the grains into the center, and the furrows that are carved in the stones' surfaces force the grains out to the edges, where the actual grinding happens."

They chose a large granite stone mill because it moves slowly, and the grain spends a long time in the mill. The oils are incorporated into the flour, which will extend the shelf life somewhat.

"Whole-grain flour tends to go rancid more quickly because of the oil in the germ," he said. "So a bigger stone will blend together everything more thoroughly and avoid some of the breakdown of larger pieces of fat."

"We can't produce white flour in this machinery," said Amber. "We have had pizza people ask for the whitest flour possible. We do special orders with special grinds for people looking for something in particular. It's an ongoing education to help people understand what the machines are and what they can do."

Michael showed the sifter, a box of blond wood that matched the mill. He told us how sifted flours are numbered for the percentage of the whole grain remaining after some bran is removed. Maine Grains makes flours that are whole wheat, and 86 percent and 75 percent extractions. He pointed to the tubes that moved grain from the bins upstairs to the equipment on this floor, the mill, dehuller, and seed cleaner. The tubes work pneumatically, and air pushes grains and flour to the gravity table and bagging machine on the ground floor. He described how the oat roller and dehuller behind him, a steely blue machine the size and shape of a furnace, worked in concert with the gravity table, sorting out small groats to make a steel-cut grind, and the smallest oats for animal feed. The oat hulls were spun out in the cyclone and sold for a variety of uses, from compost to feed.

"This is the same kind of mill that Jennifer Lapidus and Emma Zimmerman are using," said Amber. Emma was in the room. She gave a talk at the conference about her family's Arizona mill, Hayden Flour. The other mill, Carolina Ground, is in North Carolina, and Jennifer Lapidus has spoken at other Kneading Conferences. "There's a small community of people using these in the United States. When we turned ours on the first time, we got Dave Miller on the phone and held up the phone, had him listen in California. We asked, 'Does this sound right?'"

People appreciated the humor.

Before grains production was centralized, this size mill was all over the place, Michael said.

"What is this mill for?" a woman asked, pointing to a small red mill that sat opposite the big one.

"When my neighbor and I began growing wheat for my bakery, the founder of Johnny's Seeds, Rob Johnston, gave me that mill. He also allowed me to use Johnny's seed cleaning equipment to clean my wheat," Michael said. That's when he began to learn the processes that he uses as a miller. Using an 8-inch mill taught him he wanted much bigger millstones, to make better flour, but the little mill is still used for test batches.

The woman had the same mill, and was setting it up to make flour for the pizza oven she ran on a Maine farm.

As people funneled out of the mill room, a farmer who had come all the way from Missouri and wanted to start milling lingered, looking over the machines.

Down a tight hallway, we stood in front of the old control center for the jail, a narrow room with long rows of windows. Computers still sit in the space; they belong to a group of high school students that designs websites.

"When the building was up for sale and I toured it, there were still inmates," Amber said.

Someone asked about the crimes, and a man offered a quick categorization. "Drunks, drugs, and domestic violence," he said. "Standard central Maine poverty-stricken problems."

Amber might have phrased it differently, but she let his wording stand and moved on to describe how the building has become a food hub.

"The Pickup Café runs a multifarm CSA and does some wholesale distribution to institutions, like the hospital," she said. Some farms use part of the jail as a root cellar, and a basement room is set to become a cheese cave, to make good use of its granite block walls.

Amber led us toward the front door and the sunny Saturday that awaited us outside. She reminded us that the farmers market was to the right.

I walked to the café and spied Mark Fowler, asked him if we could have a cup of coffee. He was from Kansas State University's International Grains Program and had come to discuss the big picture of grain production. When he'd spoken the day before, some people had left his talk, uncomfortable, as the discussion turned to GMO crops. Big wheat, small wheat—I wanted to know how grains grow and move through the world.

Mark is a tall man. At the café, the table seemed tiny between us. He taught the grain milling course that Amber attended, and had consulted with them as she and Michael set up the machinery and flow. Mark has worked as an engineer in the milling industry, setting up operations at the other end of the size spectrum. This was his first small project, and he's helped on a few more since. He would stay a few days into the following week to work on the flow and efficiency of the mill.

The International Grains Program makes millers, training them at a small flour mill on campus. I'd been to that building, and it was bigger than a high school, but Mark Fowler meant that it was small for the industry. The Kansas Wheat Commission had its offices on campus, too, and I asked him how the university and the industry were linked. He explained the wheat check-off program that takes two cents per bushel of wheat the farmers sell, and directs the money toward marketing and research. Other states with wheat commissions use different formulas to catch and deliver funding for research and advancing the industry. He told me about origin mills, which are located near growing regions, and destination mills, near population centers.

"I know you don't have a crystal ball," I said. "But do you think we'll see more small mills like this one in the future?"

"These things are not something you can franchise," he said. Every locale varies too much for a cookie-cutter approach to work. "Takes a lot of passion from the farmer side to the consumer side. Not everyone or everywhere can make this happen. It's a real grassroots effort."

He traced the success in Skowhegan to this effort.

"The mechanics of the mill are easy to solve. The marketing on either side, that's the challenge," he said. "You need the right mix of people wanting to see their community succeed."

When we said good-bye, I stood outside the old jail and thought about what success meant. The desire for change that began the conference is now a formal nonprofit, the Maine Grain Alliance, which runs the conference and peripheral outreach projects. The MGA's mobile wood-fired oven is parked in the former exercise yard and travels to schools and events, where Breaducation bakers teach about grains, seed to pizza.

The municipal parking lot beyond the farmers market is under construction, thanks to a grant to make the lot safer for pedestrians. In that grant, the investment that had been made in the building was accepted as matching funds.

Across the road is another testament to the mill's influence. The bank was using the old Grange building for storage, and was going to tear it down for parking. Community members bought it, with the promise to put a business on the first floor. The dance floor on the second floor will be used for special events and social activities, just as Grange halls were when they were brand new.

Nearby, the Hight family has sold cars for generations. Debbie Hight is an active board member of the Maine Grain Alliance and one of the round-the-clock volunteers for the conference. Before all of this activity, she told me, there weren't many reasons for young people to come home after college. The brain drain isn't fully corked, but now, returning home is a viable economic option for some.

And what about the ones who never leave? Does all of this flutter matter? Each year when I come to Maine, I wonder how the bread bubble touches everyday life. This thing that means something to me, does it affect the average resident?

I walked around the farmers market and looked at what was for sale: blueberries, beer, lettuce, and kale. I spied Blue Ribbon Farm's pastas; the whole wheat noodles were made with flour from the mill. But other vendors made products from more common and inexpensive flour, and those breads and treats sold quickly. Seeing such flour was disappointing, but I know that's silly. The market isn't just a showcase for the mill. It is a place for fresh, local food.

People can pay for that food with cash, credit cards, or SNAP benefits. Having a democracy of payment options costs time and money, and not every farmers market makes access a priority. Some customers get free produce, and a program that funds prescriptions for fresh fruits and vegetables pays the farmers. The farmers are dispensing produce like vitamins, with a doctor's script. Participants receive counseling on healthy eating, and data on their choices are collected to report on how subsidizing consumption of healthy food works, or doesn't work.

Such measurements won't show all the benefits of the program, just as the mill's profits can't be read merely in dollars. Dollars don't account for the accomplishments I could see: the Grange, the repair to the parking lot, the young people Debbie Hight said were coming home after going away to college. These intangibles add up and increase, attracting more people to give more time and work to the MGA.

I was seeing that up close, staying at the home of board member Susan Cochran, a doctor. Rather, I was seeing her investment because I wasn't seeing her. I saw her at bedtime, and by the time I got up in the morning she was already gone, helping to steer people and things at the conference.

The benefits can be read more concretely, too, in contracts that the mill writes with farmers. From these grains, the mill makes 125 tons of stone-ground flour, and just as much dry rolled oats. Oats are generally steam-treated, which makes for a longer shelf life but reduces flavor. Maine Grains' oats travel as far as New York City and Rhode Island. Whole Foods is selling these oats and Maine Grains flours, and making a Maine loaf with flour from the Somerset Grist Mill, too. These are not ideas that are floating around, but bags of Maine wheat and oats, tagged with a processing date, and the name and location of where the grains were grown.

These products represent more than the mill. They show the result of networked efforts throughout the state. For about a decade, many organizations have been investigating grain production. MOFGA, the Maine Organic Farmers and Gardeners Association, and Organic Valley, the dairy cooperative, share a full-time employee, John Chartier, who is working in Aroostook County, the state's most northern, remote, and Iowa-like section of farmland. John's job is to help farmers develop a regional food system in an area that exports most of what it grows: potatoes for McCain's, oats for Quaker, and malting barley for Canada. The imagined transition includes farmers growing outside commodity markets, and growing for businesses like the Somerset Grist Mill. This is not a novel effort for Organic Valley; the company also helped Maine farmers start their own feed mill, Maine Organic Milling.

Ellen Mallory from the University of Maine sits on the board of the MGA. Through the University of Maine's Cooperative Extension, she organizes a grains meeting late each winter, drawing almost one hundred farmers. Some are already growing row crops and forage crops, and are curious about making grains work in their systems. Others are actively tackling grain farming and sharing their experiences.

Ellen has the slim build of a runner and the steady gaze of a keen observer. She has had her eye on wheat worlds for a long time. During her graduate work in eastern Washington State, she worked with farmers who grew for the commodity export market. Her position at UMaine is extension outreach and research in sustainable agriculture, and she coordinates field days and other efforts, like the winter meeting, to help build resources for grain growing. She has a good take on how the movement for local grains is progressing here and elsewhere, and sees the mill as an important factor in nurturing a regional grain system.

"Any new venture, any enterprise on the farm, you need to know that the market is going to be there for you to access," she said. "It takes a little bit of time to get that trust."

Farmers in Aroostook County who sell barley or oats by the truckload to Canadian markets need to see infrastructure like the Somerset Grist Mill succeed and provide a stable market. Yes, farms there are already growing for the mill, and how these relationships develop will determine if more farmers take the risk to break out of the commodity market. So far, so good. The mill is a steady customer, and increasing its purchases each year. As this pattern continues, it should nudge along production to meet the mill's demand.

Ellen is optimistic about the power of consumer interest in local foods.

"I don't think I've ever seen anything quite as promising as this focus on local, not just in terms of moving sustainable agriculture forward, but in moving agriculture forward in general," she said. "It's definitely in reaction to people feeling like something is broken in our food system. As people start going to farmers markets, they get more connected to where food comes from, and understand how that food is produced."

At first she thought the interest in local was niche and boutiquey, but the interest didn't drop even as the economy has continued to suffer after the recession. What looked like a trend has some resilience, and to her that's exciting, whether wheat reconnects people and agriculture, or any other food is the glue.

Talking with Ellen, I started to think of the conference as a sewing machine. The event stitched together professional and serious home bakers, not just to one another and this great amount of bread skills, but also to their prime ingredient, flour.

"I have a different relationship with my ingredients," said baker Ciril Hitz in his closing remarks at The Kneading Conference in 2013. "It may not be a direct communication with the farmer or miller, but everything has changed."

Ciril Hitz is a beloved instructor, teaching baking at Johnson & Wales University in Rhode Island and at his home workshop, BreadHitz. He has a friendly intensity, and has led many workshops in Maine. His awareness of farmers and millers has grown as he's listened to them discuss their work and concerns. Knowing what goes into getting flour into a bag makes him appreciate the process, and his place in the cycle.

This is the kind of connection Ellen Mallory loves to see. When she worked with growers in Washington State who sold on the export market, they had no link to the people who used their crops. Mostly, the grain went to Asian noodle markets. As a researcher, seeing how her agronomic investigations play out for growers is rewarding. Even better is bringing together farmers and bakers so they can learn about each other's work and concerns.

Ellen facilitates such connections. She partnered with colleague Heather Darby from the University of Vermont on the Northern New England Local Bread Wheat Project, a USDA OREI–funded project. The $1.3 million grant allowed them to bring farmers, millers, and bakers together in a range of ways to increase organic wheat production. In addition to researching effective fertility and weed management strategies for organic wheat, the universities tested over fifty wheat varieties at multiple locations in both states, seeking varieties with good bread-baking qualities and good growing characteristics. Bake trials held at King Arthur Flour's Norwich, Vermont, location drew people together to test the best results. The grant also sponsored trips to Denmark and Quebec to study small-scale grain growing and processing.

A key mentor in the project was Jim Amaral, baker and founder of Borealis Breads. He worked in the bake trials and helped farmers understand wheat from a baker's perspective. He had pioneered access to local grains in the 1990s; his yen for local flour helped start Aurora Mills and Farm, a whole-grain mill in Aroostook County. Matt Williams, at the time a crop specialist at UMaine Extension, began growing organic wheat for the bakery and enlisted other farmers to grow bread wheat, too, initially funneling the crop to another mill for processing. In 2002, Williams started Aurora Mills, and ten years later his daughter Sara joined the family operation, switching from a career in landscape design.

All of these efforts in grains are attracting other young farmers in the state to grow staple crops.

Adam Nordell and Johanna Davis are vegetable farmers who also grow heirloom grains. When they were choosing where to locate Songbird Farm, the projects in Skowhegan helped them pick western Maine. At farmers markets they often bring their bicycle-powered mill to lure customers to grind fat red and yellow kernels of Abenaki flint corn into cornmeal.

The first years they grew grains, they focused on corn, because they could pick it by hand with help from friends. They got access to a combine and other equipment once they met Julie Zavage, who was working at the Somerset Grist Mill. The three worked together to grow more and other grains, starting a dry goods CSA. They also sold their beans and heritage corns and wheats through farmers markets and co-ops. Julie has left the mill and Maine, but Adam and Johanna are buying her equipment and a farm. (They had been leasing land.)

Other new farmers are focusing entirely on grains, like Sam Mudge. Apprenticing at a dairy that grew grains to supplement the cows' ration got him into grains. He started Grange Corner Farm in 2010 and focuses on grains at a very small scale, under 20 acres. At first Sam and his partner sold cornmeal through word of mouth, and then they branched into farmers markets. They applied for a technical assistance grant through the Maine Grain Alliance, and used the money to design a logo for selling grains at farmers markets. Grange Corner Farm sells rye, cornmeal, and wheat flours, featuring the new homey but professional-looking logo. The farm also grows and sells raspberries and strawberries.

A self-described seed nerd, Sam is growing out einkorn, emmer, and other heritage grains for another MGA project, multiplying rare and heritage seeds to meet consumer demand. Sam had already been growing out seeds he loved himself, so he was a perfect grower for the program.

"I think the university research has really gotten the word out to the general farming community about bringing that part of farming back to the Northeast," Sam said. The Canadian Maritime provinces were already doing research on grain growing before the University of Maine and the University of Vermont got involved, and he said it's encouraging to have the universities circulating information through bulletins and crop reports.

Like Songbird Farm, Grange Corner has had difficulty sourcing equipment for growing grains at this scale. Sam has a combine with a 10-foot head, and he bought a mill to be able to make flour on his own schedule. He is always on the lookout for more, and reads the classified ads in farming newspapers, searches online and at auctions, and asks around, eager to find what other farmers use and what gems might be hiding forgotten in barns.

To me, his scouting for equipment parallels the community prowl for grains in and around Maine. People are thinking about how to grow and use them, trying new methods and unearthing old tricks. People are working alone and coming together to share what they've learned. Grains and bread are building a mesh of connections. Commerce is shrinking down to a community level, bringing more people and products eye-to-eye. Bakers are learning what farmers need, and vice versa.

Dusty Dowse is a good example of someone engaged in that community prowl. He's a biology professor at UMaine, and an MGA board member who coordinates baking classes at the Somerset Grist Mill's classroom. He covers lots of ground at The Kneading Conference. One year he led a class about the history of corn, emphasizing the agricultural origins and use by Native Americans. The class was not just heady but also hands-on, and people learned how to nixtamalize corn for tortillas. (Nixtamalization is a process of soaking and cooking grains to break down cell walls; in corn, using an alkaline solution like lye, this results in *masa harina*.) Another time he led a discussion of gluten and celiac sensitivity with Dr. Susan Cochran.

In 2013, at the beginning of the conference, Dusty made his standard announcement that people who wanted continuing education credits should see him, and he also made a request. He invited bakers to help make bread for the Common Ground Fair in September. He recruited a crew of ten and got two ovens. I didn't get to see the setup, but Dusty said it looked like a medieval village bakery. His crew made 1,000 pounds of bread for the volunteers who run the fair. To Dusty, and to Albie Barden, that bread was a realization of the impulse that started the grains work they'd undertaken. They had gathered around a wish and made it happen.

I hated to leave Skowhegan. There was something there I wanted to hold and know, and when I finally pulled myself away, I felt like I was swallowing a fist. This year, when I got to my car, I found a gift. Someone had tucked a paper bag into the crack of my opened window. The loaf was dense, almost like a cooked pudding. I tore off chunks as I drove out of town, loving the moist darkness of spelt, rye, and buckwheat kernels. This was more real than most food. Maybe that was because I was so hungry for what I was leaving, the nebulous something that had been made. Was I leaving connections? Community? I was so disoriented that I had to keep checking my phone to make sure I was actually going home, not driving in circles.

Around me, the gas stations and houses fell away. Fields and trees took over the landscape. A ring of clouds edged the horizon, same as the year before. Those clouds felt like a hug, a good-bye from the unnamed many who make the conference, and the mill, tick and tock and go.

Another image struck me as I ate that lusciously heavy bread. When the conference had ended, Susan Cochran took a bunch of the work-study students out to Lake George for a twilight swim. This is a big lake, surrounded by tall trees. The setting seems epic, because nature is so much more prominent than us toothpicky observers.

The long day was ending, and we were the only people at the small beach. It was chilly. Susan is a lanky woman and she loves to swim. She led us into the lake, promising there was a rock where we could stand. I swam a little, and then stopped because my arms hurt.

"We're almost there," she called. Her voice is always smiling. She is someone who is convinced that every minute has its own adventure.

"Come on," she encouraged the swimmers, pulling people along with her delight. "I think it's over here. We're almost there."

CHAPTER SIX

EAT THE LANDSCAPE

*P*ancakes have caught and held my attention for forty years. My father was the Saturday-morning pancake maker, and he invited me to help when I was very young. I remember staring at the inaugural pancake, a single disk on the cast-aluminum griddle. I was looking for bubbles to form and break so I could flip the cake.

I still stare at the griddle with the same readiness and awe. The red box of Aunt Jemima or Hungry Man mix is not on the counter. The blue-and-white-flecked bowl is filled with mixes we make at home, and my awe for pancakes has grown as I've learned of the people and labor that go into making flour. I can easily slip into a grain-based reverie any morning, one hand on my hip, the other hoisting the spatula in the air like an extended index finger. I worry that my predilections foist too much of my favorite food on my family, so I started asking Felix, my youngest, what he wanted for breakfast. He always says pancakes, which thrills and relieves me. When my older son requested a lunch of crepes for his sixteenth birthday, I was pleased.

Francis lives at boarding school. We brought crepe batter, and cheese and greens for fillings, for a picnic. We carried our kit to the cow pasture near his campus. The boys climbed a birch tree, perching on its ridiculously thick branch, to chat and wait and watch. We never see cows in this slightly hilled field in western Massachusetts, just some of their remnants. As Jack set up the portable stove, I stared at the kids, long and lean as branches.

They are inquisitive little old men who are most content in nature, although Felix would prefer an audience to receive his many words. Francis is not shy of talking, but he is a great audience to the world.

Once the griddle was hot, I poured the batter, made from rye flour, and watched the crepe, waiting for the surface to turn from shiny to matte.

The attention I have to pay each crepe or pancake gives me plenty of time to consider everything, including my ingredients. I love this forced meditation. That day in the cow pasture, I wondered about my affection for the flour. Did I love it so much because I could trace its path? I couldn't remember many, many things about Francis, but I can trace our path. I recalled the oddity and wonder of having another person inside me. His beautiful, pickled face at birth. The way he stood and gobbled a roast chicken at a coffee table on his first birthday, eyes wide with delight at what he was eating. I don't love him just because I know where he's been, but our memories sure do seal him to me.

Now that the picnic is a memory, crepes seem the perfect salute. He is six feet tall, and growing into his life with others. He had a girlfriend this year, and there is something new about the way he hugs me. Other people are loving his skin. He is a crepe, wrapping up the world and devouring it. His eyes are wide with discovery.

Crepes are like skin, supple and thin, slippery from heat and butter. I love how they take up the whole griddle, and then the whole plate. Rolling them up with my fingers, they feel like skin, too, warm and moist from the shower. Inside, I tuck a swipe of yogurt. For lunch, I add beet and wheat berry salad, or cucumber and yogurt with herring. The skin analogy suffers once the crepe is tubed over a lumpy filling, but, putting a bite in my mouth, there it is again: a slippery skin of flour giving me a kiss.

This is a new joy. Until recently, I had no need for crepes because I don't like eggs. I love their function in baked goods, but I don't like the taste. Plus, I inherited a knee-jerk disdain for French food as part of a classbound rejection of anything fancy.

I never made crepes until a friend said she was having trouble with flour she'd gotten at a farmers market in New York. I knew that the flour was from Farmer Ground or Wild Hive, and I had to defend it.

The problem, Leah said, was that the batter wouldn't hold together. When she made the same recipe using whole wheat flour from the bulk bins at the coop, the crepes didn't fall apart.

Luckily, I'd just read an article about how to make good crepes. Stuck without a book in the doctor's office, I was a captive audience. Even if crepes were too fancy for the likes of me, and eggy enough to stay off my plate, they were griddle food. Maybe I could learn something to apply to pancakes. The author said to mix the batter thoroughly, in a blender if possible, and then let it sit for at least half an hour. Don't put too much batter on the griddle. Do slick that griddle well with butter.

I applied these rules to Leah's 3–2–1 formula—3 eggs, 2 cups milk, and 1 cup flour. We made two batches, one with that anonymous co-op flour, and one from Farmer Ground whole wheat pastry flour. We didn't have a blender but we did have a whisk, and we let the batters sit as long as we could. There were a lot of hungry people, so there was no way we could give it half an hour, but we gave them 10 minutes.

Making something you don't love is an odd endeavor. Usually I think my way into a food. I need to know something well before I bake it, to estimate an outcome. I can taste my pancakes in my head before I make them. I know how the pancake should be on my tongue, dense but not tight, a short, soft cake slightly sweet with whole wheat and malt.

But with crepes, my only navigational tool was knowing the flour. I certainly had no fondness for the eggs. As a kid I skipped sleepovers where the parents insisted on big egg breakfasts. The crepes I was making would probably be too eggy for me, but eating wasn't the point. I just had to make the flour work.

The stakes were high. My friend took a risk and bought local flour, and was disappointed. The flour was expensive. The feel-good stuff—supporting New York State farming and products—doesn't matter if the functionality is off. Why pay extra for something that didn't behave as expected? Since I knew the people behind the flour, I felt an extra layer of responsibility. If I couldn't get the flour to perform, I'd be betraying them and their work. I wanted my friends to admire the flour, or at the very least not think ill of it.

I want to see small flour mills survive. There are not many of them, and the doubt about local flour is huge. People think that wheat doesn't grow outside the grain belt. Isn't flour just powdered cardboard, something to add water to as you're building bread?

We don't grow oranges or bananas in the Northeast, so why grow bread? Professional bakers protest that local flour is too variable.

I never had trouble with what I make. The pancakes always work, and so do the pies and cakes. I do not bake much bread, but we eat plenty of homemade pizza, and our crusts are never rejected. If I were operating at a commercial scale, and feeding more than my family, who are used to a steady diet of whole grains, the parameters might be different.

Whole-grain flour, with its pieces of bran, works differently than white flour. I suspect that some of the hesitation about using local flours is related to the inclusion of bran. Stone milling, which is a more affordable entry point than roller milling, uses all of the grain: the bran, endosperm, and germ. Even if bakeries are familiar with whole-grain flours, they might know roller-milled products, which don't have the germ and might have smaller bran particles.

Regionally grown and milled flour takes interpretation because it is different. That is part of the point. The goal of any localized food production is not to replicate the anonymity available from commodity products. The flour is less standardized than national brands, though, and the variability has given it a bad reputation. Professional bakers balk because they don't have time for another set of variables. Local flour, or fresh flour, or whole-grain flour doesn't always do what they expect when they're making dough. Wrong turns at a bakery level are expensive.

Wrong turns used to be common at home, too. Throughout the nineteenth and early twentieth centuries, American bakers had to be savvy about flour. What you bought—or ground from your own grains—could range widely in quality and performance. Cookbooks reflect how much people needed to know about baking and ingredients.

A pamphlet of Pillsbury recipes put together by Fannie Farmer and two lesser-known cooks is a good window on other flours at the turn of the century. The company was not just bragging as it warned, "These recipes are for Pillsbury products only and are not intended for flour and cereals of inferior quality and strength."

Okay, maybe this was bragging, but not without cause. When stone mills were the primary route to flour, the wheat generally came from nearby. There is nothing wrong with proximity, but relying on what your region produced increased the likelihood of irregularity in flour. Quality varied from year to year depending on the weather as the wheat grew; quality also changed over time during storage. Good flour depended on a good growing season and a timely harvest. Keeping grains dry and free from insects, rodents, and other spoilers between harvests was key, too. Any number of problems could make flour less than ideal. Poorly dressed millstones had an impact, and, depending on a miller's system, you might be going home with flour made from someone else's grain, anyway.

The pamphlet, published sometime between 1889 and 1909, was an ad for a new style and brand of flour, Pillsbury's Best. I love how the writing and recipes address an audience that knows flour:

> *PILLSBURY'S BEST BREAD*
> *To one quart of milk or water add one cake of compressed yeast; add flour to the thickness of batter, and let it rise over night; then add enough flour to knead softly twenty minutes, or until it will not cling to the board, as it requires more kneading than winter wheat flour; let it rise in the pan, then make it into small loaves, and let it rise again. Bake in a moderate oven, and do not let it stand in oven after the bread is done.*

Given that Fannie Farmer is known for standardizing measurements, the unspecified amount of flour is curious. Most intriguing, though, is that the recipe refers to winter wheat, and how the new flour would require more kneading.

The switch from winter to spring wheat was revolutionary, because it absorbed liquids so differently. To me, the bigger difference is that this flour was blended from many, many fields.

> *The best wheat is grown in the Red River Valley of the North, Minnesota and the Dakotas. It is known as Hard Spring Wheat. It contains more gluten, more phosphates, more health-giving and strength-sustaining qualities, than any other wheat on earth. Pillsbury's Best Flour is made from this wheat. The wheat in over three hundred elevators and stations is analyzed, and only such as comes up to the Pillsbury standard is used for Pillsbury's Best Flour. The remainder is sold. Pillsbury's Best Flour is made from the choicest of the best wheat in the world.*

This was a curated mixture. Though surely not the first time a mill pooled grains, the scope of the pooling was much broader. The Pillsbury-Washburn mill was the first large roller mill in the country when it was built in 1881. Shortly afterward, this and other Minneapolis mills began blending from wheat from hundreds of farms, a feat that was only possible because of developments in transportation and equipment that enabled handling such quantities, both on a farm and at a mill. Milling at a community scale meant taking what the season offered. This mill could be selective because it had more choices, and because those choices could be defined.

Being able to test and analyze grains for elements that affected baking quality gave millers the capacity to blend grains from different fields and farms, and create a more uniform product. Historically and currently, variations in soil and climate conditions affect how grains grow. During the heyday of Genesee wheat, bakers across the Atlantic loved the new flour because it absorbed water differently. The spring wheats in Pillsbury's Best Flour also absorbed water differently, and the booklet cites the fact that bakers had to use less of it in recipes as a money-saving advantage. Professional and home bakers were responsive to the behavior of this ingredient.

A century after consumers were introduced to more homogeneous flours, the terrain of baking has completely changed. Until the 1920s, most bread in America was produced at home, and some cookbooks from that era specify soft flour, meaning pastry flour, in recipes for quick breads, cakes, and piecrusts. Home bakers now might choose cake flour, but I bet that information on flours is off most people's radar, unless they are avoiding wheat or gluten.

Most people think that flour is flour. I don't blame them. They haven't had the chance to see flour speak for itself, let alone express originality. White flour doesn't have much flavor. If you can't taste flour, why would it matter what kind you use?

Flour's job in baking is mostly structural, holding up ingredients that deliver taste, like butter, sugar, and chocolate. Other local foods make easy sense. Tomatoes are a no-brainer. Meat and eggs are relatively easy sells, too, thanks to awareness of the ills of factory farming. Starches don't seem as essential to buy off-grid. Potatoes, rice, wheat—we eat a lot of these, and they are generally bland and cheap. The volume factor works against grains because commodity pricing assigns insignificance. Most people have limited food dollars to spend, and by dint of cost, starches fall to the bottom of the list of things that are important to buy nearby.

Another strike against local grains, in terms of romancing the consumer, is the obscurity of the producer and the production process. A combine is a combine is a combine—what's the difference if it runs somewhere in the anonymous middle of the United States or somewhere in the invisible middle of New York State? Paintings of harvest scenes have a pastoral appeal, but grain is never going to charm the pants off the general public. Not like the idea of cattle chewing grass and never hitting the feedlot. Flour is safe from the sentimental realm. Unless you are me and have a magnetic attachment to the stuff.

That attachment can wield some power. The crepes I made with the doubted flour were preferred over the store-bought stuff, and we made an extra round. I was relieved. Reading that recipe at the doctor's office helped me interpret the flour. Determination and a dash of griddle confidence were also at work.

Confidence and fondness won't make flour work, however. Since the shift is not lateral, consumers need help adjusting to flours. Mills that have a teaching space host classes. Farm stands offer recipes and websites that guide people through the process of baking with novel flours. Larger regional mills post flour specification sheets online. For professional and serious home bakers, these numbers show how flours fit within a set of expected standards. Protein content and falling number are really important to leavened doughs, and millers blend to hit the marks that fall in bakers' comfort zones.

Millers buy grain that has been tested for protein levels, moisture content, test weight, and DON, which is an acronym for toxins that result from a fungus that can grow on grain. Sound wheat will weigh about 56 pounds to the bushel, and have certain moisture and protein levels. DON levels have to be below 1 part per million. The falling number, which shows the enzyme activity in flour, should be greater than 250.

If this sounds complex, it is. Making flour is a nuanced process, and requires grain to meet a set of givens for milling; those parameters are further narrowed by what bakers can use.

Grain reaches physiological maturity before it is dry enough to safely store. Moisture levels of 13 percent and under are necessary for proper grain storage. Higher than that, the grain can mold in storage and is more susceptible to pests. Protein levels vary according to grain varieties, season, and farming habits. Soft wheats have lower protein contents, around 9 percent;

hard wheats tend to have protein contents that are higher, and can climb up to 14 or 15 percent. In arid regions, protein contents are higher in bread wheats.

DON stands for "deoxynivalenol," a mycotoxin produced by the *Fusarium* head blight fungus. Ingesting DON causes vomiting, which is why the condition is also called vomitoxin. The fungus comes from the environment, via crop scraps like corn and wheat stubble; spores of the fungus can also reside in the soil, or travel by wind. When grains are in the flowering stage, the fungus can enter the plant and begin to affect kernels, bleaching them and sometimes causing them to turn pink. These scabby kernels can be cleaned out of the grain, so millers can accept grains with DON levels slightly higher than the 1 ppm that is acceptable for the final product.

Falling number is a measure of how fast a plunger falls through a slurry of flour and water in a test tube. Failing this odd exam means that too much sprouting occurred before a field was harvested. When seeds start to grow, the starches that bakers rely upon to feed fermentation are damaged. Some adjustments can be made to compensate for high falling numbers, such as adding malted barley flour to the wheat, but there is a target range, and if the falling number is below 250 the grain probably won't make the grade for flour.

Protein levels indicate what quantities of gluten-forming proteins—gliadin and glutenin—are in a flour. Acceptable protein levels vary according to the end product and the style of bakery. Large bakeries that "bake by the numbers," using flours that hit anticipated protein levels and other specifications, usually can't tolerate the irregularities of flour from a small mill. The production volume is too great, and employees might not have the necessary time or skills to pay attention to how flour behaves in a dough as it mixes and rises, adjusting ingredients or fermentation times accordingly. However, the most important part of a bakery's readiness to use local flour is the baker's own attitude or commitment.

Some bakers are more open than others to the differences that local flour presents. Peter Endriss is very ready to meet a flour at the bowl.

I met his bread six months before I met him, at a tasting of regional flours. His rye, dark and dense, sweet and sour, sat in my brain like a gargoyle perched on a building. I made a note of his name, and followed his progress from selling at farmers markets in New York City to opening a restaurant, Runner & Stone, in Brooklyn. The bakery is a star part of the operation.

"The only non-local flour we're using is artisan white bread flour from Central Milling," said Peter one afternoon, joining me in a lull between the lunch rush and cocktail crowd. The tables were empty, and so was the space at the bar where the breads usually sit. Good press has helped the loaves march out the door long before lunch, so Peter gave me a verbal tour of the invisible breads.

Runner & Stone features baguettes that are white, whole wheat, and buckwheat. It also makes a whole wheat walnut levain, Bolzano rye, sesame semolina, and a rye ciabatta, all with varying percentages of whole-grain Farmer Ground flours. The brioche and croissants have 10 percent whole wheat flour. Champlain Valley Milling provides the white spelt flour used in the bakery's pretzels, which are modeled after a southern German pretzel that uses Dinkel flour, which is also spelt. These breads are built with many qualities in mind.

"First, I want the bread to be nice," he said. I'm not sure what nice bread means to this baker, but I know he is very nice. Years ago, he met me for coffee just because I emailed him and asked. He might not extend the same generosity to every fan, but this seems emblematic. He is always accessible, even if his bread vanishes.

Peter has a big smile. He maintains eye contact when you're talking with him, and if he can't ignore his phone, he excuses himself. He is small, wears glasses, and keeps his dark hair cut close to his head. His pacing is very friendly. Whether he's chatting, baking, or doing a presentation, he is not in a rush. He lets everyone and everything have time, even—or especially—flour.

"I think the pre-fermentation of whole-grain flours is key to taming them," he said. "Throwing a fistful of whole wheat flour into the dough is like throwing in a bunch of razors. If you pre-ferment, it just kind of softens everything rather than interrupts it."

Pre-ferments are common in artisan baking because they develop flavors and give flour, even bran-free white, a chance to hydrate without salt, and sometimes without leavening. Peter was already building doughs with this method before he applied it to whole-grain flours. Classes at Camp Bread and Wheat Stalk, the Bread Bakers Guild of America's forums, and recipes with multiple ferments suggested this route. A big influence also came from working in a bakery just outside Paris for a month.

"The bakers would grab bread out of the oven and everyone would break it open and smell it and eat it. The bakery was very, very focused on the flavor of the product and less on the appearance," said Peter. "Their product and their approach to baking was very inspirational to me."

For instance, L'Etoile du Berger used a levain—a liquid sourdough—in its croissant dough. At Runner & Stone, the croissants use a levain, too, to build structure and flavor. At first, the levain was made with white flour, and whole wheat flour was added in the final mix, but now the sourdough starter for the croissants is fed whole-grain flour, too.

Runner & Stone opened in December 2012, but Peter was baking for farmers markets under that name before then. He met June Russell when she spoke about the burgeoning interest in regional grains at a Slow Food event in 2010. "I'm Peter Endriss and I'm going to be your best friend," June recalls him saying. He is an ally to the movement because he's committed to using locally grown grains. His skills elevate that commitment beyond good politics to the realm of great bread.

The week I visited him, the *New York Times* had praised five loaves in the city, and two of them were his. The fact that all the breads were made using regional flour was not mentioned. This different, in the best sense, ingredient didn't yet register in the narrative of good bread. The bakery is meeting some high expectations, as is the restaurant. That press, however, does point the finger to quality ingredients, speaking the farm-to-table language. The flavor potential of flour is not as well understood. Bread like this can help change that.

"Very rarely do we get asked where our flour comes from," said Peter, noting that white breads like the baguettes are the biggest sellers. When questions arise about the flour—and he did encounter more of them when selling at farmers markets—they focus on how much whole-grain flour is used in each loaf.

Peter and his business partner, chef Chris Pizzulli, wanted to use as much local flour as possible at the restaurant. While roller-milled local flour is available, Peter was more inclined to use stone-ground.

"I have a degree in environmental science. I think that plays into it," Peter said. He worked as a civil engineer before switching to a career in food. "I think my experiences in studying natural resource management and doing fieldwork associated with that [give] the farm a stronger presence in my mind when I look at an ingredient."

He likes to use a variety of grains, to help counter the tendency toward monocropping, but Runner & Stone's breads don't carry the heavy halo or punishingly self-righteous textures associated with the 1970s era of whole-grain breads. His interests are decidedly flavor-forward.

"How much whole grain can we add to a baguette and still have it be my impression of a baguette?" he said of the calculations he makes. This means a thin crust and an open interior with a flavor that is not too sour; something that is pleasant to eat and a little lighter than a whole wheat sourdough. "I could make an 80 percent whole wheat sourdough, but then it wouldn't be a baguette."

Peter's first job baking was at Amy's Bread. In addition to the stage in France, he also spent time at a bakery in his father's hometown in southern Germany. He was head baker at Per Se and helped develop the bread program for Per Se and at Thomas Keller's bakery, Bouchon Bakery in New York. He worked at Hot Bread Kitchen while baking for farmers markets. These experiences gave him a good idea of what kind and size of bakery he wanted to operate. He wasn't interested in running a retail bakery in New York; these operations generally rely on hefty wholesale accounts as well, to balance the budget. The profit margins on a retail bakery are too thin to stand alone. Pairing with a restaurant and bar means the bread doesn't need to support as much overhead as it would in other sales venues.

This scale is also an assurance of good communication. Peter works with only four bakers. If something goes wrong, the small staff can talk it over with one another and trace the problem. He's worked with one of the bakers for many years, starting at Amy's Bread.

"Akemi and I are like four arms with one body when we're in the kitchen together and know what questions to ask each other to keep the process on the rails," he said. The bakery's size means they can bake by feel, rather than just formulas. All bakeries use recipes (formulas) and weigh ingredients. Smaller ones can be flexible to the whims of a living process, and accommodate variations in dough, whether caused by humidity, measuring differences, or inconsistencies in flour.

"If you have 300 kilos of baguette dough going down the track and it's off by 5 percent, it's going to be a big deal for you. If our 5 kilos of dough is fermenting a little too fast, we just put the tub in the fridge and fix it," said Peter. When I asked what would be required of regional flours for greater adoption, he said, "I wouldn't put the onus on the flours. I put the onus on the baker. I think the flour is fine."

This confidence in flour came across when he taught a class in local flour baking at The Kneading Conference. His goal was to show how simple it is to incorporate local whole-grain flours into production.

"I'm going to hand-mix a spelt ciabatta that's 85 percent whole-grain spelt," Peter told the group of mostly professional bakers. "The key to working with local stone-ground grains is just giving the dough a little more time."

Peter then went on to explain the dough and its flours: Farmer Ground Flour spelt and 15 percent white organic bread flour from Central Milling in Utah.

"This is a pretty coarse flour," he said of the spelt. "People might be afraid to use a high percentage, but it has really nice fermentation properties, and it absorbs really well. It gives you nice structure in the end." He noted that the higher hydrations possible with this kind of flour are an economic advantage. That hydration would prevent the bread from achieving a lot of volume, however.

"I think that maybe people need to adjust their expectations of what bread should be," he suggested. "Maybe bread shouldn't always be super puffy, or have shapes that make the most beautiful supermarket sandwich."

The mechanics of using the flour was only an entry point to discussing other barriers that can make shifting to regional flours difficult, like availability and cost. Nestling the bakery inside a restaurant, Peter said, was part of his solution to the problem of cost, not just the cost of local flours but also that of running a bakery.

"We're a bakery and a restaurant, and our food costs at any given plate per ingredient would be 30 to 40 percent. About a third of the cost of that salad is what we would pay for ingredients," Peter said. He pays about $23 for a 25-pound bag of organic white flour from the Midwest, and about $30 for a bag of whole-grain organic New York State flour. Conventional flour costs about $13 per bag.

"For bread, the cost of ingredients is typically 5 to 10 percent or lower if you're using conventional wheat, so maybe it goes up to 15 percent for organic and local," Peter said, "but it's still in the scheme of what food should cost."

"This flour costs $1 a pound, which I don't mind paying," another baker said. Getting people to pay for that increase, however, is tough, when the perception is that bread should be inexpensive. Or that it mysteriously arrives at the table in a restaurant, like silverware and water.

"Part of the learning curve is getting people to pay for it, and also, I think, questioning the size of some of our bakeries," said Peter. Some bakeries that are thought of as artisan have gotten so large that their processing is mechanized, which makes these types of doughs impossible to achieve. In this sense, size limits skills. Reskilling the workforce and training people for baking as a career, Peter said, is something that belongs in the conversation about changing flours.

Toward the end of the session, Trine Hahnemann, a chef and baker from Denmark who was also presenting at the conference, raised her hand and spoke.

"I think we are a bit ahead of what's going on here," she said. Denmark has been working with local flour for the last twenty years. Her catering company serves lunch for two thousand people every day, and all the bread is homemade from local flour.

"We don't have bakers. We have chefs who bake," Trine said, using her experience with spelt to address the broader issue of local flour. "I don't think we're ever going to get the same spelt. I've been working with it for more than ten years and every batch that comes in is different. We have to get to know it every time. That's also what makes the excitement, makes the craftsmanship. When a new spelt is coming in, we don't know how long the resting period is going to be. We know we're going to have to work with it. We can't just say bread flour is the same as all the ingredients we work with. I don't believe it's ever going to be the same, and I love that—it's exciting. Sometimes your bread is just going to go *pfft*, and all our customers joke about the dough didn't rise but it still tastes good."

People laughed and Trine paused.

"We need to reeducate our customers," she said, not in a hopeful, wistful way, but with conviction. Customers see that the bread doesn't look like the bread from a week ago, so they get an understanding. A few people applauded.

"The other thing I want to say is price. I don't think we should negotiate with the farmers. I think it's really, really important that we negotiate with the customers. That is our role, to say this is what it costs," Trine said. Lots of people applauded. "In the future, this will be more and more expensive because of water, because of climate change and a lot of other things, so that's our responsibility. Because if you can't make a living farming, what's the point?"

Peter finished his demonstration, dividing the dough and showing the strong structure that developed in the ciabatta as it rested. People asked about his using olive oil, and he talked more about the flours he used.

I was optimistic about the reality that Trine presented. I don't think that we can get the same acceptance of novel flours as in Denmark, mostly because America is so big and unwieldy. We are a nation of resettled rebels, and our proudest bread banner is Wonder Bread. Northern Europe has a strong tradition of whole-grain baking and cooking. Our next step with grains won't look like what happened in Denmark, but their model is encouraging nonetheless. And every model helps imagine another new thing. Change is one thing we Americans love. Maybe that can work to our advantage.

Other New York City bakers are also using local flour. Zach Golper of Bien Cuit in Brooklyn is using North Country Farms flour. Austin Hall, whom Peter worked with shoulder-to-shoulder at Per Se, is baking with Farmer Ground Flour at She Wolf Bakery. The bakery supplies Andrew Tarlow's restaurants: Reynard, Marlow & Sons, Diner, Roman's, the bar Achilles Heel, and the store Marlow & Daughters. She Wolf began at Roman's, where Austin Hall baked bread in the wood-fired pizza oven. In January 2013, the bakery moved to rented space in a shared kitchen and began baking seven days a week. The bread also enjoyed great press, and with good reason. The whole wheat miche—a kind of French country loaf that might be the poster child for the artisan bread movement—is still sitting in my mind, staring at me like Peter Endriss's rye gargoyle.

For these urban bakers, freshness is significant. Flour from nearby has very active enzymes, making it great for sourdough fermentations. Zach Golper said that he could see the difference in the fresh flour as soon as he put it to work. Its scent and lively activity were wildly exciting. By the time the bread was baking, his enthusiasm was at a fever pitch, baffling some of his workers. They couldn't understand how flour could be so evocative.

These flours are just as active on the baker's imagination, serving as a bridge between farm and city. That link matters to Zach, too, who took a tour in the summer with his daughter to see mills and grain fields in Quebec and the Finger Lakes. Most eaters won't notice the link between field and loaf—not now. But maybe in a decade, the public consciousness will have shifted and more people will be thinking backward, from baguette to bowl to mill, all the way back to the seeds.

The concept of farm to table is more than a passing infatuation; it's a recognition that eating is an agricultural act, and an attempt to combat some of the many problems in that process. One measure of economic success in America is housing starts, but such metrics aren't going to keep their meaning as wage depression continues to affect our country. If New York City needs New York flour, maybe some of the pressures of suburban development can be eased. These ideas and the concrete actions taken toward change are invisibly buttering our local bread. And bakers are leading the way, helping us to eat, and reconnect with, the landscape.

CHAPTER SEVEN

EVERYDAY REVOLUTIONS

*L*ucky Dog Organic is a farm in Hamden, New York, a one-road town in the Catskills. Richard Giles and Holley White run a storefront farm café that kisses that road, and own or lease 150 acres of land in the area. Between 40 and 60 acres are in vegetables any given year, and about 25 are in grains. The food they grow goes to the shelves in the farm store, area farmers markets, and to New York City, at the Union Square and Fort Greene Greenmarkets.

The couple left the city and began farming almost fifteen years ago. The grains are a relatively new venture but echo Richard's other, much different farming experiences, growing commodity grains in the South.

"We grew hard winter wheat. I can't remember the varieties," Richard said, clipping the words in his southern accent. Each word is more tight than drawn out, but his pacing is easy and open. There's a spaciousness that comes when talking with him, room for thinking. "These were strictly commodities. A crop that you take to the elevator and dump it, and take what they offer."

His father worked at an experiment station on the Mississippi Delta, and Richard worked farms in eastern Mississippi and western Alabama. When the farm he was managing sold, he went to grad school for fiction writing at the University of Alabama. He moved to New York City in 1995, and met Holley there in 1999. Shortly after they met, they left the city for the farm.

Farming up north, he's steered away from commodity crops because they were a dead end. Lucky Dog has focused on specialty vegetables, and this niche found a receptive audience, especially among the restaurant chefs and buyers who frequent the Greenmarket. He and Holley have experimented with different sales formats to make the financials of farming work. More wholesaling, then less. More farmers markets, then fewer. For a while, they ran a CSA, and this is what led to grains.

Michael O'Malley and Mercedes Teixido built a home a few miles from the farm. They teach art at Pomona College in Southern California, and come to New York for summers and breaks. They joined the CSA and developed a friendship with Richard and Holley.

In 2011, when Michael was on sabbatical, he lived in the Catskills. Surrounded by farms, he realized he wanted to grow wheat. This is not your average thought for anyone, let alone an artist, but bread had crept into his art and life.

Good bread first hit him in Maine, where he was working in ceramics and eating loaves from Bodacious Breads. (This bakery is now called Borealis Breads, and has been a leader in using regional grains.) Though Michael vowed then to learn to bake, it wasn't until he was living in Pasadena, where he couldn't buy a decent loaf of bread without effort, that the decision became a necessity. This coincided with a trip to Barcelona to research the sculptor Gaudi. After he ate exceptional pizza from a wood-fired oven, building an oven made it onto his list of musts as well. First, he dove into baking. The habit fed a lot of people and projects.

"I always bake two loaves and give one away, so it serves as this kind of bridge between myself and the people in my life," said Michael, who races through words and ideas. He made a sourdough starter, too, and always gave that away. Soon he started teaching people how to bake. He also began making temporary ovens as art installations, and baking was always a part of these art pieces. Growing grains was the obvious next step.

Michael bought a combine on Craigslist, a little tow-behind All-Crop, the classic harvester coveted by people who are not going to grow many grains. Once he had the machine, he talked to Richard about growing wheat for bread. He knew that Richard grew rye as a cover crop, so he figured he'd have a ready partner for the project. Since Michael had the harvesting equipment, and was willing to find and pay for seed, as well as work out storage in the barn, Richard was game. In the back of his mind, he'd wanted to do something with grains, but keeping his eye on so many dozens of vegetables, the most he'd ever done was harvest some rye from time to time for seed.

"Winter rye is so vigorous, but we're real busy in summer with the vegetables, so I tended to plow it in," Richard said.

The grains have injected some extra enthusiasm into his farming work, which is often monotonous and hard. Richard also enjoys the angle that growing grain brings to something he knew in an entirely different context.

"Our approaches are so different. He's a baker. My experience is with conventional grains, farmed as a commodity crop that has a narrow life," said Richard.

To come back to grains with Michael's insight is nice, and more akin to the way the farm handles vegetables, as foods that have value and variety.

The collaboration made the endeavor possible. The first year, in the fall of 2011, they planted Arapahoe wheat and some rye. Michael fixed up the combine and readied a storage bin upstairs in the barn. When they harvested 500 bushels, both were surprised. This was something real, not conceptual.

"The aroma, the feel, the freshness, you never expect that from flour," Richard said when they had Farmer Ground Flour, which is a couple of hours away, grind some Arapahoe.

The informal partnership gave Richard and Michael more than either could have achieved working alone. Lucky Dog had a new, intriguing product, and Richard had gained a new perspective on grains. Michael could get to know flour from the ground up, and gain more understanding of a substance and process that fascinated him.

"Honey, cider, bread . . . I think they're some of the real mystery foods of our planet. It was a phenomenal experience seeing what we grew. You don't have quite that experience with a tomato," said Michael. You can't, because you can just pick a tomato and put a little red or golden sun in your mouth. For bread, you need people, tools, and time. Consumption requires a lot of processing and attention.

To Michael, the orchestration of flour, water, salt, and wild yeasts parallels the practice of making art. Both start with a vision and use materials to get to an end that is shared with an audience. Whether people are viewing art or eating bread, artists and bakers are connecting them to manifestations of ideas and labor. The materials act as a translator between people and things. Getting involved with wheat added layers to Michael's engagement with his artistic process.

In the fall of 2012, they planted 12 acres of grains, and the following July harvested about 1,200 pounds an acre. The pair made their first shared investment, a small Clipper air-screen cleaner, and again had Farmer Ground Flour mill some of the harvest. Holley uses some of the flour for breads she bakes at the store. Most of the grains and flour go to the farmers market at Union Square, where Richard gets to see how they are received.

Lucky Dog has a wide and discerning crowd shopping its tables. Its vegetables are known for quality and appreciated by chefs who shop directly at Greenmarket. One restaurant sends around a chef that Richard thinks of as a kind of vegetable bouncer, because he won't take anything less than the best. Many shoppers have a more pedestrian attitude but equally strong attachments to good food. These people are not wealthy, but come from pure food traditions and still prioritize fresh foods. Most customers are open to trying whatever he brings, whether it's watercress or Arapahoe flour.

More than the flour, though, people are excited to try the wheat berries, using them for sprouting or in stews. Greenmarket uses his grains in demonstrations, promoting them as it would any other product. Recipe cards help sell the grains, and it doesn't hurt if Richard has a story about making a risotto from wheat berries with peas and carrots.

Richard still plants rye as a cover crop, building organic matter in fields that grow vegetables. Each year he bumps up the acreage in grain. Michael has a 20-inch Meadows mill, but has been too busy to set it up. The two want to mill at the farm or nearby.

"If we could mill we could sell more. Right now wheat berries sell more than flour, but it's very low volume and I would like for it to become something wholesale," said Richard. Not wholesale to a grain elevator, but specialty wholesale, maybe to breweries or distilleries. Right now, the grains don't net enough income to warrant extra labor. "It's still this thing that can't quite get done because it's a secondary business for both of us."

The dimensions of the grain enterprise, though limited, are satisfying, and contribute to the farm's viability. Lucky Dog is one of many start-up farms that are recarving an abandoned path to New York City markets. Railroads once helped develop this area as a dairying region, and people shipped fresh milk and cheese to the city. Today, though, parts of the

Catskills feel like farm cemeteries. A historic marker between the interstate and Lucky Dog, for instance, notes the site of the first commercial pasteurization in the United States.

This is still farm country, though. A nearby high school has one of the last vocational-technical farm training programs in the state. Fields that aren't hayed are planted to corn or soy. Conventional dairies face markets that favor large herds. Forty years ago, two hundred cows constituted a big dairy, but now that's considered a tiny herd. Currently, fluid milk prices are strong, but this stability exists within a precarious price system.

Lucky Dog is distributing for other farmers to New York, offering space on its truck to a few producers who set simple pickup spots with restaurant clients. A wave of small artisan cheesemakers is popping up. Some are conventional dairy operations wanting to diversify that see cheese as a value-added product; others are start-up operations created by non-farming folks who got a taste for good food and more tactile, rewarding careers.

The Catskills, because of their proximity to New York, have had the chance to be a lot of things to a lot of people over the years. Some railroads just served excursions, toting people to panoramic vistas and long-porched hotels. On walks in the woods, rusted car hulls and plow handles dot the trails, the same as mossy rocks. Humans have been here a long time, trying to figure out how to live. Nature is trying to absorb the remnants of different attempts, like the shells of Borscht Belt hotels aching their way into the ground. Arrowheads are easy to find in the many thready streams whose names are mash-ups of Dutch interpretations of Native American words and *kill*, the word for "creek."

The mountains are modest, pocketing those creeks and settlements. Everything is tight. Paved roads fade fast to dirt. The fields and lakes kiss each other, and the diminutive mountains pucker up to the sky. When it thunders in the Catskills, it sounds like overgrown imps are bowling in heaven, and that heaven is in reach, just the other side of the horizon.

Some communities are picture-perfect, ready for centerfolds in home and garden magazines. Colonized by second homeowners from New York City, each Greek Revival house is perfectly painted, every porch colonnade restored. Gardens are planned bursts of height and color.

Other towns are more standard upstate blends of hard times and modest survival. Houses in decent shape sit next to dilapidated trailer homes bleeding rust at the seams. There are stark contrasts between the haves who come to the Catskills for fresh air, and the residents, both longtime and more recently settled, who are trying to make the beautiful environment work for them. For their part, Michael and Richard would like to see their joint work feather out and create more opportunities.

"A couple of farmers around here have heard about the project and are curious. That's hopeful to me," said Michael.

People are watching, but old-school farmers are not yet leaping into the act. Artisan cheese has a track record of success, and organic vegetables seem to be working well, too. But grains are a high-volume game. Lucky Dog sells whole grains and flours for $2.50 a pound at markets, and though that's more than the price of commodity organic equivalents, it won't pay for extra help, let alone infrastructure like a dedicated mill room, to expand production. When they do set up the mill, perhaps it could be a community operation.

In the meantime, Michael's connections to bread and wheat are many in California, too. His LA drive times include hauling a mobile oven to different sites.

The Michael O'Malley Mobile Oven, or MOMO, for short, is the nick-name of the oven he built. He creates events that are part art installation and part social experiment—opportunities for people to walk that tightrope between nature and culture, art and life, and participate in something that has a lineage and heritage but feels brand new.

Baking pizza and bread in public makes visible a process that is often hidden. The oven began as a way for the sculptor to complete a thought about wheat and bread, but the oven is a starting place, not a stopping point. There are more requests for MOMO than Michael can fit inside his full-time teaching schedule.

"I just keep getting blown away by how many people want to learn how to make bread," Michael said. Blown away, yes, but he sees the logic of the interest, too. "In a highly mediated and controlled world, people are looking to exercise their agency in small ways. So having a garden, baking bread, making your own cheese, whatever it is, gives you this agency and sense of determination of who you are."

Choices allow us to alter the current homogenization of life and work. Intellectually, he sees the DIY impulse in Gandhian terms. Gandhi got the British out of India by creating mechanisms for people to determine their futures. Cottage industries threw a wrench in the machine of colonialism. In much the same way, the re-skillings people pursue are tools that can throw wrenches in the machine of capitalism. And Michael sees bread as a particularly useful tool for setting new patterns.

"Bread hits you on all senses, backdoors your brain with this smell, taste, texture, crumb, sound," he said. "This is why teaching people how to make bread is important. The world is only going to change by all of us incrementally reimagining the world where we live."

A new mill in Pasadena is a platform for such reinvention. The oven and Michael travel regularly to Grist & Toll, the first new mill in the Los Angeles area for more than a century.

When the mill opened in November 2013, MOMO came to help celebrate flour. Baker-owners Nan Kohler and Marti Noxon wanted the allure of live fire, and the communion of breaking bread. Members of the LABB, Los Angeles Bread Bakers group, brought dough, and Michael baked the loaves. Even though cooling is the final stage of baking bread, deemed essential to fully developing flavors, no one waited to cut those loaves, some of which had flour Nan milled at home.

People brought crackers, cookies, and cakes made with that flour, too. These delicious messengers of the flavors of fresh flour disappeared quickly. All kinds of people came to wish the mill well: friends and family, baking historians, Janice Cooper from the California Wheat Commission, and Glenn Roberts from Anson Mills in South Carolina. As the Osttiroler ground its first flour, people stared and applauded. A few of us got close, climbed on a ladder, and looked at the berries in the hopper. Down on the ground, we put our hands in the new flour. The novelty was striking. The mill, pretty as a piece of furniture, is as confusing as the Wizard of Oz. We have no frame of reference for milling, not even words in a song. Wheat has "amber waves of grain" to keep the concept of grain growing somewhat alive. Silhouettes of wheat heads garnish bakery logos. General Mills is the name of a cereal company, but what exactly is a mill? The idea is generally lost. Millstones, the only associated object, prop mailboxes on suburban lawns.

Nan wants to put the visuals back in circulation and make milling normal again. This is a bold thing. The idea of an urban mill, or any mill, is hard to conjure. Her beautiful pine-planked Osttiroler mill is a curiosity, sure as any wonder P. T. Barnum found and touted. Through the ambassadorship of baking, Nan is installing the tools of milling, and the loveliness of fresh flour, in everyday life.

Grist & Toll sits in a long-blocked grid of big anonymous and windowless buildings. The streets are flanked with palm trees, and the area is zoned for industry. For the most part, the only way to tell what is made in each structure is by the labels on cargo vans removing goods. A few blocks away, Jones Coffee Roasters carved out a path for craft manufacturing with a retail outlet.

I like the hint that coffee roasting lends milling. Thirty years ago, the shiny copper bellies of coffee roasters helped people learn about good coffee. Now many cities have a café with an on-site roaster. Maybe thirty years from now, we will see small mills everywhere. Sounds optimistic and absurd, but I want flour to fit back inside our lives. That's why I'm crazy about Grist & Toll.

This urban mill started outside a restaurant in Paris. Nan and Marti were strangers, waiting to be seated, and said hello. They kept bumping into each other on the trip, and carried the friendship home, united by an easy rapport and passion for baking. Marti Noxon is a screenwriter with credits like *Buffy the Vampire Slayer*, and each year she undertakes a study of a single theme. The year of pie. The year of bread. Nan regrets missing the sweet fallout she could have enjoyed as Marti baked her way through the year of cake. But she's had no shortage of great baked goods in her life.

For years, Nan baked intensely and recreationally, too. A wine seller, she unwound from work in the kitchen. While she loved all kinds of cooking, the desserts she made got noticed, and people asked Nan to bring them to dinner parties and gatherings. Her work in wine covered such a large territory that eventually she felt like she was living at the airport. Needing a change, she started baking for a farmers market, and her relaxation became her profession. That leap led to another, and soon she was baking for a restaurant.

But Nan wanted to push her baking one step farther, and a PBS show gave her the lightbulb moment for the mill.

In an episode of *Adventures with Ruth*, food writer and personality Ruth Reichl visits baker Richard Bertinet in Bath, England, and together they visit the miller who grinds the flour. The baker says what he'd like to make and the miller says what he's milling, and the decision of what to bake is based on their exchange, which includes a discussion of the climate that day, and the kind of growing season the grain had.

"Why don't we have that?" Nan asked herself. "The food culture is on fire in the United States. There's so much talent, yet there's no fresh flour."

Oddly enough, Nan had seen the episode more than once, but the mill only jumped out at her when she was scanning the world for clues about how to expand the business of baking. Marti was helping her brainstorm, and encouraged her to dive into the project.

Figuring that milling must be ridiculously hard or impossible, Nan started researching. Once she found that wheat was available, and that other mills existed, there was no turning back. Tenacity has proved a good partner.

"Historically, mills were the epicenter of developing cities," said Nan, but mills and farming gradually moved out. Like urban farms, she wanted to be in the thick of things, to be a hub for the kind of exchanges necessary to integrate fresh flour into modern kitchens.

As Nan scouted locations, reception for the idea ran hot and cold. City managers bent over backward and opened up their zoning codes, but licensing entities couldn't grasp what the milling would look like. Yes, there are farms in California that mill, and bakers who mill, but for the most part these are add-ons to existing operations, and the extra equipment slid in with little regard, like another mixer or tractor.

So industrial mills became the reference point. How great was the risk of explosions from flammable dust? One authority thought a sloped floor and drainage system would make sense for a mill, even though grain and flour need to be kept bone-dry. Nan had to start from scratch with each potential location and set of inspectors. She even sent a video of a small mill in action to one health department, trying to demystify the process.

While negotiating for a space, Nan taught herself to mill. She bought a set of 12-inch stones cut by Roger and Larry Jansen in Chico, California, and put it in her garage. Milling days, she worked in the driveway, experimenting with grains and grinding styles. Since she couldn't use all she milled, and wanted more feedback than she could get in her own kitchen, she gave flour away to home bakers.

If people asked what to make, she suggested they try shortbreads, which are a great vehicle to test-drive most flours. Odd textures or protein levels might trip up the leavening in cakes or breads. Besides, butter is a convincing delivery system.

"Butter and salt go on steroids in shortbread," Nan said, somehow amplifying the flour's flavor rather than drowning it. Everything tastes richer, more interesting and complex. Describing shortbread she brought to a holiday party, she really riffed.

"We used soft white, and the hard red added like a base note," she said. "There was this depth you didn't have with regular white flour. Using fresh flour, the chocolate you add shines more, the butter tastes more buttery and toasty."

Nan's life in wine gives her a good vocabulary for discussing flavors. The dictionary of fresh flour is still being written, and she loves watching people try to articulate what they taste, trying to tag flavors and characteristics.

"People can't quite put their finger on it," Nan said. "Like a lot of other things, tasting is when you start developing a palate. It is going to be a while before we're there with wheat."

The mill really puts together everything she's done and loved: her skills in baking, the curiosity that makes her a talented educator, and her intuition in sales. From home bakers to the more rarefied world of restaurants and chefs, she knows how to discuss a new product, and has very specific conversations about flour based on who's going to use it. This is what she did with wine. She was able to speak to everyone, from people who would only drink burgundy or Bordeaux to those who just wanted a really delicious bottle of wine on the table.

Discussing flour with bakers, she shapes the message around flavor and performance. For instance, Sonora wheat is lower in protein and absorbs liquid very differently than standard flour. Nan can tell people how to use it in piecrusts and artisan bread.

The deeper she gets with flour, the more ridiculous it seems that this basic food is so late to the great ingredient party. Even if the omission makes sense.

"I don't even want to tell you what I paid for the right chocolate and European butter, but no one has been looking at flour?" she said. "For so many years, flour has just been a filler, something that gives body and structure to your cookies and cakes. We haven't had the ability to think of it as a texture building block or flavor building block until very recently."

Alternative grains are relative new kids on the block, and alternatives to mainstream milling are following suit. As we reshuffle our relationships to grains, education is a key component. Grist & Toll creates chances for that to happen, like hosting bakes with Michael O'Malley and MOMO. Once a month, people sign up for bake times and juggle their dough schedules to drive and walk bread to the oven. The scene is a good time for bakers to get technical information and troubleshoot problems, or just enjoy one another's company.

Nan also coordinates with small local producers to sell bread and pastries on Saturday mornings. People can preorder online from rotating menus focused on the mill's flours. Pastries come from Sugarbloom Bakery, and Bread Culture makes bread.

***

Bread Culture is a model of how many things—health, nutrition, and minds—can change when you switch your approach to flour and bread. Mary Parr works from her home kitchen, operating under a cottage food license. Her relationship to bread is a compelling story of reform. She was pretty sick the last few years, suffering from digestive pain. Tests for celiac disease came up negative, but doctors suggested she might find relief by going gluten-free. She swore off bread and gluten, and this, plus acupuncture, somewhat curbed her symptoms.

"I still had stomach pain. I still had to watch every bite," she said. She was always worrying about when or if her stomach would feel full of pieces of glass.

Mary is studying Chinese medicine, and when she was headed to France for her honeymoon, a friend at school kept telling her to try some bread.

"I was able to eat everything there," Mary said. "I came back and thought there has to be a way to change this for people. Gluten isn't an enemy."

Mary used herself as an experiment, trying to bake bread she could eat. She took a class with Michael O'Malley and started using wild yeasts and long fermentations. After eight months, her digestion has drastically improved. She still can't eat conventional wheat products, like commercial crackers or cookies made with supermarket flour, but she really feels the difference from eating her own sourdough bread.

"I won't use anything else but Nan's flour," she said. She loves the feel of the flour. Having her hands in the dough, she feels creative again, which is a nice break from the scientific rigors of her master's program.

Mary thinks the flour is miles away from any alternatives. Freshly milled flours are not oxidized as much as others, and the nutrition is more bioavailable. Another important factor is the sourdough cultures. She believes the probiotics in the bread are a salve. When she runs out of bread and doesn't eat it for a few days, her stomach woes creep back.

Mary has the zealotry of the healed. Having found a solution, she wants to share her success. She began bringing bread to school simply because she loved baking and had more than she could eat. Once she discovered she could get a cottage food license, she began selling at school, and at Grist & Toll.

Having been mired in big questions about gluten, she is sensitive to the issue. She makes a gluten-free loaf with a gluten-free starter produced from teff flour. She loves the feeling of sharing bread with people who have had to avoid it.

When other people find they can eat wheat bread again, too, it reinforces her belief that all the fuss about gluten is not really about gluten.

"It's about how we live and how fast we live," said Mary. Scaling down and slowing down milling, and giving dough a long time to ferment and break down gluten, is critical, she feels, to reviving our relationships to bread and other things. She sees the panic against bread as part of a larger problem. "We are not eating food anymore," she said.

Mary bakes about twenty loaves a week for others, and once she graduates she plans to divide her time between baking and an acupuncture practice.

Renegade baking like hers is lively in the area. The LABB has a thousand members, many of them perhaps inspired to figure out bread as Michael O'Malley did, because LA has a dearth of good bakeries. One of these bakers, Mark Stambler, is responsible for legislation that enabled people like Mary to bake from home.

Mark has a wood-fired oven in his backyard, and a passion for baking that took off into a small, but illegal, career. His bread is so good that the *Los Angeles Times* got wind of it. At first, the writer agreed to write about him without identifying where he sold bread. However, the editor wouldn't allow that. After the story ran in 2011, the health department forced him to stop selling bread. With the help of the Sustainable Economies Law Center, Mark did the research and legwork for the California Homemade Food Act, which has been a boon to small food businesses. In the first two years of the law, more than four hundred licenses have been issued in Los Angeles County alone; thousands have been issued throughout the state.

Now fully legal, Mark bakes bread using flour he mills himself. He calls one loaf the LA Miche, which contains wheat grown in Santa Barbara County. He found this wheat through Nan and Grist & Toll.

Home bakers have been a great source of support for the mill. Word of mouth leads people to the flour, and then the flour creates its devotees. Long known as the best advertising money can't buy, word of mouth is fed by the educational opportunities Nan creates at and beyond the mill.

Grist & Toll is not just a new business but a new product whose nearest equivalent is considered bland and unimportant. Nan has to fight that impression, and she hosts classes to introduce the flour through a range of bakers and baked goods. Variations on the theme of bread are sell-outs. San Francisco's bread all-star Josey Baker came to the mill with his new bread book. Jonathan Bethony from The Bread Lab at Washington State University brought his snappy, thoughtful take on sourdough. Don Guerra from Barrio Breads in Arizona left a buzz, too.

Nan has taught classes on single topics like pies and cookies. She does outreach off-site, teaching at places like the Institute of Domestic Technology, a grown-up twist on high school home economics courses.

The institute is the brainchild of Joseph Shuldiner, who describes himself as a lapsed graphic designer and art director. The courses tap into an appetite for re-skilling that seems at odds with the freeway lifestyle, but perfectly jibes with the flour mill and the alternative routes to bread that Mary Parr, Mark Stambler, and Michael O'Malley are making.

The Institute of Domestic Technology started in Altadena, where the lots are big and the land is fertile. As in other parts of the country, a homesteading wave took hold in the first decade of the twenty-first century, inspired by food safety recalls, movies like *Food, Inc.*, and writers like Marion Nestle. High food prices and job slashing during the recession nudged along self-sufficiency as well. People were questioning food, and their answers set off more questions.

"People were raising chickens and goats in their backyard," Joseph said. "They wanted to know how to preserve the excess vegetables from their gardens and make cheese from goats' milk."

That's when Joseph's curation instincts kicked in. He started a series of demos and showcases that grew into the institute and a full slate of year-round classes about base ingredients and food handling. One of the most popular workshops is a full weekend focused on baking bread. Writer and urban homesteader Erik Knudsen takes people through Volkenbrot recipes and overnight ferments, and Joseph teaches simpler basics like crackers and tortillas.

Nan's classes on flour fit right into these anatomy studies of food. She brings grain and flour samples, and uses a tabletop mill to grind flour on the spot. Through hands-on contact with fresh flour, Nan gives people words for a familiar but entirely unknown food.

Joseph and Nan have team-taught at Grist & Toll. Using recipes from his cookbook *Pure Vegan*, they swapped out whole grains and flours. Nan talked about how to think about whole grains and how to approach them as ingredients. They showed how to adjust recipes for different kinds of flours and grains. Students were jazzed about the class, and also about being so close to the mill. People could see the warehouse, stacked with bags and totes of grain. The visuals are, as Nan predicted, a mighty tool as she tries to install fresh flour in people's lives.

While time-consuming, education might be the easiest part of her undertaking. She is writing the book for regional grain flow in Southern California, which means there's no model to follow. She is not aligned with farmers, as some start-up mills are, so sourcing grains has been tricky. The state grows more than a million tons of wheat, but much of it is exported. Large-scale growers and grain brokers don't want to deal with a small mill. The logistics don't make sense. No silos? A parking lot without enough room for a tractor-trailer? Her work looks Lilliputian.

With smaller farms, the question of who should pay for quality testing is up in the air. When wheat comes in full of weed seeds, and it has, the nearest grain cleaning facility is more than four hours away.

The glamour of heritage grains and the presumed badness of modern wheats is an even larger problem. This creates a dichotomy and hierarchy that makes the revolution she envisions impossible. A middle ground is necessary for all strata of consumers to have, as she described in a blog post, "superior flour, responsibly grown and milled locally, and have access to it more than once a week or once a month."

This won't happen without an acknowledgment of the limits of land-race and heirloom varieties, Nan believes. Modern plant breeding is not poisonous, and it can help sustainable farming. Better wheat doesn't necessarily mean antique wheat. The economics of growing these storied grains is tough. Some farmers who plant heritage grains want to capture all the value of the crop by milling it themselves. Though that's understandable, Nan feels this stance overlooks the skills of milling, and the necessity of having a place to learn about flour. More than other raw ingredients, flour needs interpretation, at the mill and in the bowl. An egg or an almond doesn't ask too much of a cook, but flour can miserably fail.

The frustrations are plenty, but for Nan the rewards are greater. A new farmer found her through the Internet, looking for a place to sell the Joaquin Oro—a type of hard white wheat—she'd grown. Another wheat she's thrilled to use is named for its color, Charcoal. The gluten properties make it impossible for bread, but for pasta, she said, it's drop-dead gorgeous. Nan has cold-smoked wheat prior to milling, and is looking forward to working with black barley. These oddities might make her sound like the Willy Wonka of flour, but most of her work is not grandstanding, just standard. Fresh flour is crazy good on its own, and it is charming new fans on a regular basis.

"Every day the best people on earth walk through my doors," said Nan. "They say, 'How are you doing? We want to make sure you're always here. We've been telling everyone about you.' They hang out, ask each other questions."

These are home bakers who already love baking. That they can love the experience even more is very gratifying. This is what Nan wanted: to connect people with high-quality expressions of a presumably banal ingredient.

In olden days, when people brought grains to the mill, standing in line was a great place to catch up on gossip and news. The news at Grist & Toll is the flour itself, and Nan Kohler is making stellar flours that deserve a lot of discussion. She's also blazing trails in a profession that I hope will become common in my lifetime.

CHAPTER EIGHT

TRANSFORMATION

*C*hange in agriculture doesn't happen overnight, but for twenty-five years Les Fermes Longpres has been redefining itself in Quebec, just west of Montreal. Most recently, the farm's definition stretched to include my favorite thing, flour milling.

Loic Dewavrin returned to the family farm in 1990, when his father decided to divide the land between him and his brothers, who were already farming.

"The revenue was not sufficient for all of us, so we had to find alternatives to what we were doing," said Loic, who left a career as an industrial engineer. Conventional corn and soy were the farm's products, and the brothers needed to increase the worth of what they produced.

"Since the beginning we were looking for a project to be able to transform the production in order to reach the consumer," said Loic, his English nicely twisted up in a French accent. The first project they undertook was pressing the sunflower seeds they grew to produce oil.

Value-added is the American term for adding value to farm products, either on-farm or within a locale. *La transformation* translates loosely to "processing" or "value-added." However, Julie Dawson, who has worked with farmers who bake and mill in France, says the American equivalents don't capture what the French term implies. The way that bakers talk about craft and artisanal skills gets at the sentiment better. *Transformation industriel* refers to industrial processing, so, depending on the context, the implications might vanish.

Though Loic doesn't use the words *transform* or *transformation* with drama, the relationship between farmer and goods sounds more affectionate. To me, it shows the attachment that the term *value-added* misses. The practice is critical for farms to succeed outside of volume sales, and there's no shortage of connection between jam makers and their berries, or cheesemakers and their cows, yet *value-added* sounds clinical, not close. *Transform* and *transformation* are more representative of the connection.

The sunflower oil sold well, but the plan was never to rely on one crop, and this particular crop was tricky to grow. The process of making the oil, however, is exemplary of the family farm's explorations as it transitions from conventional crops to ones that offer more economic independence.

In a broad sense, farms are in a constant state of revision. Large or small, farms are not factories with set flows of materials through machinery. Each season is an opportunity to try a new tool or tactic. Farmers are a breed of innovators, thinking on their feet because they often have to make do with what's on hand. However, when Loic moved home, the normal work-in-progress status of the farm shifted to a radical project as the brothers chose to switch from conventional to organic production, and move out of commodity sales.

As an industrial engineer, Loic was trained to have his eye on systems and solutions. On the farm, Loic and his two brothers, and now his son, investigate an element and innovate handling procedures from the ground forward. Everything from seed procurement to soil composition came, and still comes, under scrutiny.

"We are trying to find the best combination of operations in the field to respect the environment as much as possible," said Loic.

The farm is about 1,500 acres. Converting from commodity corn-soy sales wasn't a simple matter of monetization, but a path to farm health and self-sufficiency. Loic suspects this quest for autonomy was inherited.

"During the Second World War, my father moved to a small farm in occupied France. Up to sixty people were living on the farm. This was only sixty-five years ago, and while many were starving, they were able to survive without depending on the external system, which had failed," Loic told a group of farmers and food activists in Vermont in early 2012.

The Northern Grain Growers Association had invited Loic to be the keynote speaker at its annual conference. At the time, I couldn't see why seed saving mattered. Two years into following flour, I still didn't realize the importance of seeds. I followed wheat back to the field but remained focused on the products of farming, not the process. What Loic said didn't penetrate my ghettoized consumption-based thinking, but I dutifully took notes.

"When we transitioned to organic, it was important to minimize input expenses and, also, maintain our seed supply," he said. "When you produce your own seeds, you get good-quality seeds, without foreign weeds, and vigorous seeds without disease."

Seed producers can't afford to just pick the best, but farmers can. Loic advised farmers to set aside the best-looking fields and crops for seed. He showed slides of the equipment they used and tweaked to fit their needs, and talked about the philosophies behind saving rather than buying seeds.

"We need seeds adapted to our soils and conditions," he said. "Seeds are like immigrants. It takes a while for immigrants to become integrated to a new place. Why not do the same for plants?"

Aside from these reasons, GMO contamination was a consideration in corn and soy (GMO wheat is not yet on the market so it wasn't an issue), as was the general issue of seed control. In Quebec, you have to fight for the right to use your own seeds because of the government's certification rules. The rules are different elsewhere in Canada, but in Europe farmers face even stricter regulations on seed sale and use. A farmer-led coalition formed in France in 2003 to fight seed monopolies and the erosion of plant biodiversity, Reseau Semences Paysannes, or the Peasant Seeds Network. This network/movement was an inspiration.

"We thought we could do the same in Quebec," he said. "This kind of living seed conservation would be better than seed banks to promote biodiversity and combat consolidation of the seed industry."

Toward that end, and to address other needs created by farming outside the dominant system, he and his brothers formed a cooperative, Agrobio Quebec. The group began in 2006, with five farms sharing ideas and resources. There are now twenty-four farm members, all organic grain producers working together to secure a good seed supply and develop marketing opportunities. The cooperation simplifies sales efforts that are necessary to breaking free from the commodity production chain. Working together, the members can reduce the costs of GMO testing and also research seed varieties suited to organics.

Farmers are always investigating good varieties, and as Loic and his family switched to organic production, building soil health was a primary part of that search. They've identified varieties suited to their land and climate, while also developing the various types of soil on the land they farm.

The main principle of organics is to start with the health of the soil. This means more than just not adding chemical fertilizers; it implies actively cultivating the soil structure and health, as much as you cultivate the health of plants. The short-term goals of having healthy and profitable crops cannot be managed without serving longer-term goals for the soil. All farmers know that the soil is their medium, their foundation, and as such needs care and attention.

"We invest a lot of time and money in improving soil structure, especially because the rainfalls we have had in the last few years are more concentrated," Loic said.

Les Fermes Longpres' crop rotations are generally four years long, cycling through soy, corn, wheat, oats, sunflower seeds, peas, and cover crops. But organic customs are not the limits of what they do.

"We are trying to find the best combination of operations in the field to respect the environment as much as possible. We do some leveling, and some tile drainage. We are using more and more green manures to fertilize our soil and maintain it in good health," said Loic. Subsoiling, or plowing beneath the depth of traditional plowing, is another method they use. This frees space for roots to grow deeper, a priority as our changing climate presents more forceful but intermittent rains. Reducing tillage is also an objective, though, to minimize disturbance to the structure of soil.

In general, *tillage* means field prep and crop handling with plows and other equipment, but in organics the term often refers to a tractor attachment's mechanical battle with weeds. The problem with tillage is it promotes soil erosion, structural degradation, and loss of organic matter. In conventional farming, no-till or low-till options rely on chemicals to control weeds. Loic and his brothers are among a wave of farmers trying to incorporate low- and no-till methods into organics.

In the midst of all these changes, eventually the family had time to pay attention to another transformation project. Together, wheat and soy constitute about 75 percent of their production, so each deserved consideration. Organic soybeans are easier to market than organic wheat, however, because the crop commands a steadier, sturdier price.

"We thought it could be a good idea to see what was possible with the wheat," said Loic. Four years ago, they began surveying the flour market.

The region already had a few small stone mills, and one somewhat larger small-scale mill, Milanaise, that uses both roller mills and stones to make white and whole-meal flours.

Interviewing bakers in Quebec, the family saw that the demand was still good for white flour. Settling on roller milling, though, they found themselves in a bind. Modern roller mills are built for high-capacity production, not farms seeking self-reliance. The solution lay in the recycling of equipment, which suited their inclination. Repurposing existing materials is how they'd tackled other infrastructure needs, from making the oilseed press to building a seed cleaning facility. After all, tinkering and time are more available than capital, and more desirable than debt.

"We didn't need a big installation. We wanted to do something at a reasonable cost because we are just a farm, and we don't have the financial capacity of a big industry," said Loic. In France, with the help of an acquaintance who became an adviser, he was able to find some suitable equipment at abandoned mills. Using a combination of old milling machines from about 1950, and new equipment for sifting, they put together a design.

The project took three years: a year to construct a new building, a long winter to disassemble and reassemble the machines, and another winter to install the system.

"We had to refurbish the machines because they were not used for a few years. We had to sheet-metal them and remount all the equipment," said Loic.

A roller mill is really a series of mills, made up of pairs of corrugated steel rollers that shear grain kernels into pieces. Think of wringer washers and you won't be too far off. Each set of rollers, or stand, is covered and contained. This new-old mill has six stands of rollers. The wheat is separated into many components as it is milled, and the miller chooses which parts to combine for the final flour. Because of the separations, roller mills make whiter flour than stone milling can, even with sifting.

Modern roller mills run at 800 or 900 rpm, but this mill runs very slowly, at about 300 rpm. The slower speed is less efficient, but this is not a race. Loic believes the slower speed creates a higher-quality product. At this rate, and through choices made in sifting and recombination, they are able to make finely milled white flours.

Distribution in Canada began in the fall of 2014, and samples of the flour are now trickling south to bakers in New England. For the time being, the mill will transform only their own grains, but eventually the mill may handle wheat from other farmers in the cooperative.

"We've got time and we want to do things correctly," said Loic. "We want to do a small quantity and then increase. It's going to take a few years to reach the entire capacity of the mill."

The mill is designed to make 1,000 tons of flour a year, but might make only 300 tons the first year as they refine operations. These are very small amounts in the white flour market. Around 300,000 tons a year is more the norm, and even much smaller mills, like Milanaise and Champlain Valley Milling, are running more grain than this mill ever will.

"I think we are probably the smallest roller mill in the region," said Loic. Indeed, this is a distinction, and one they don't plan to shake. The idea is to integrate another economic activity onto the farm. This is just one part of a diverse system, so even though flour is a volume game, they can set the numbers they want to play.

The choices they've made allow them to create something unique. Minimizing expenditures on equipment and performing most of the labor themselves, they can be selective and patient. They do not have to sell flour to everyone.

"You can cultivate relationships and make a network that is going to be what you want it to be," said Loic.

Some of those relationships come from south of the border. Randy George is from Red Hen Baking Company in Montpelier, Vermont. He often wears a bike racing cap, and I've never seen him without a smile. He visited the mill before it was open and took an interest in the flour. Randy is very creative about sourcing, hunting for the best flours near and far.

"With Randy, I developed a contact that was interesting, and that was exactly what I was looking to develop, this kind of relationship," said Loic. "The idea with this project is to stick with a few bakers, because I'm not going to be able to supply a lot of bakers."

Randy used some of Loic's flour in a workshop on bake-testing local flours at Wheatstalk, the Bread Bakers Guild of America's conference. He was very happy with the creamy color and texture of the baguettes, and began baking all of the bakery's baguettes from Moulin des Cedres' flour, as the mill is named. Aside from the flour's performance, the prospect of working with an on-farm mill appealed to Randy because it affords a greater chance of constancy in quality and supply.

"One of the problems of the larger millers is that it's very difficult for them to have something uniform over the year," said Loic. "That won't be my problem because I want to transform the wheat from the region here, the same variety, and the same environment."

That wheat will be the farm's own, grown from two varieties of hard red spring wheat that have good baking qualities and hit protein levels between 11 and 13 percent. They are trying to find new types, working on a participatory wheat breeding project with the Bauta Family Initiative on Canadian Seed Security. Through this project, Ann Kirk from the University of Manitoba is working with farmers all over Canada, planting on-farm test plots, having the farmers select varieties they like, and making crosses. For Loic, this is a wheat mission, but other farms are exploring other crops.

Similar plant breeding investigations, such as the one Julie Dawson worked on in France and those undertaken by farmers and researchers in Vermont, inspired this work. One of those farmers is Jack Lazor.

Jack Lazor has been visiting Quebec longer than Loic has been farming. Jack has a longish white beard, which is fitting for his reputation as a grains guru in the Northeast. He crossed the border early and often in his own farming career, looking for equipment and advice.

Jack and his wife, Anne, went back to the land in Vermont's Northeast Kingdom in 1975. Neither one had farming experience, so they borrowed from the past. They met while working at Old Sturbridge Village, a living history museum in Sturbridge, Massachusetts, that focuses on mid-nineteenth-century New England. Jack had just graduated from Tufts, where he wrote his thesis on colonial farming. As Anne finished up college in Madison, Wisconsin, Jack watched Amish farmers stook grain and milk their cows by hand into cans. Starting a life together in Vermont, Anne and Jack took cues from what they'd seen.

"We had to do everything the old-fashioned way first," Jack said. "I had to go through all the agricultural history in person to get to the modern machinery, but at least I knew how the old stuff worked."

Some of that equipment came from Canada. Right over the border was grain country, thousands of acres of barley and oats. Ages ago, Vermont's

Champlain Valley was the Oat Basket of New England, feeding horses on farms and in cities before the Erie Canal peeled grain production farther west. In Canada, Jack toured a still-vibrant grain culture and saw that growing grain in a northern climate was indeed possible.

"Every town had a mill, and every single mill had a seed cleaner so a farmer could bring their seeds," Jack said. "Some places had little burr mills for cracking corn. Every place had a hammer mill."

Jack fitted his farm with old machines from Quebec and befriended farmers who shared what they knew, from good seed varieties to tricks for cultivation, harvest, and storage. These farmers became his mentors, and Jack carries with him at least a generation of insight on grain farming— maybe more. Putting this information into practice, he's simultaneously dispersing it through many channels. He shares what he knows in everyday encounters with the people who have worked on his farm, and those who have moved on to start their own. He welcomes people to call on him for help, and is involved with more formal gestures too, speaking at conferences and helping to create the Northern Grain Growers Association (NGGA). Most recently, he wrote a book, *The Organic Grain Grower*, which has become a bible for home and farm-scale grain growers in the region.

The book explains how to and how not to grow grains in New England, and includes an inventory of all his mistakes and explorations. The voices of his mentors are alive in the book, which reads like the work of a mature writer, even though it was the first big thing he'd written since his thesis. He plainly wrote about what he'd lived, and what he'd learned from others as homesteading became full-time farming. Keeping a cow and growing grains for themselves led the Lazors to keeping a herd and building a small dairy processing facility, which is now tucked next to a granary inside a wizened-looking barn.

Butterworks Farm sits at quite a remove. The first time I visited, I wasn't quite sure I was headed somewhere. I climbed dirt roads and saw less and less evidence of people, emerging on a windy vastness from what had been a pleasantly warm summer day. On the top of nowhere, there was a house, and a big barn with a cupola. I put on a sweater and looked around for the farm's name. There it was, on a delivery truck. The barn's boards had never been painted, staunchly announcing age and experience in salt-and-pepper tones. The building seemed an appropriate home for the elder statesman of grains in New England.

The place is an odd nexus to me, so remote yet so connected. Butterworks Farms yogurt and kefir flow out to New York and New England, and Jack's wisdom and experience flow much farther, evidence of the strength of human connections.

Jack is a gregarious fellow, and has that kind of openness I saw in Randy George. These are people who look you in the eye like they really want to connect. So many people working in grains have this ready, sharing attitude. They are excited, like a kid with an inchworm. Can you remember that feeling, of needing to show someone a marvel? The whisper-like pinch of minuscule suction cups on your skin? That excitement of holding your finger up in the air, staring at the eerily translucent bug, a bulgy thread of pulsing Jell-O? Showing the beauty first to yourself and then saying, *Hey, look at this!*

This sentiment shines through Jack whether you meet him in person or read his book. He gets great energy from engaging with others, even as health issues have forced him to stop working. He has prostate cancer, and didn't pursue Western medical interventions. This led to his kidneys shutting down, and now he's on dialysis five days a week. Anne administers the treatments at home, and he spends hours looking at the landscape, the nearby hayfields and the distant Green Mountains. He's very grateful for the new lease on life, a lease that comes with guaranteed time for contemplation. Watching a machine clean your blood makes you look at other intersections of nature and technology. Same as a kid with an inchworm, he holds up his wonder and ponders broader concepts of health.

"Those 6 to 8 inches of topsoil, that's all we've got to sustain us," he told me on a cold day in his warm house. Spring had officially sprung, but Vermont took no notice, and the snow was getting a fresh coat of snow. We ate beef and wheat berry stew that I brought, and toast with the farm's butter.

"Commercial agriculture is so profit-driven," Jack said. "We're just taking carbon out of the soil. I'm at this point right now where if the farming I'm doing isn't making the soil any better, then maybe I better be doing something else."

Cash-cropping grain was never a main motivation at Butterworks, but it has figured into operations, partially because grain production of any scale requires infrastructure, and that scale is larger than what a farm and family can use. The acres planted to grains crept up to 225, and a portion of that was always dedicated to growing feed for the dairy cows. The granary has a 24-inch stone mill to make whole wheat flour, and a gravity table for seed cleaning and sorting. People sit and sort black beans in an upstairs room with a great view. Yet as the demand for food-grade grains increases, Jack's heart is with the soil and other grasses than grains.

He wants to grow nutrient-dense forages and feeds, and minimize the grain needs of the herd. He doesn't regret what he's given to grains. The pursuit fed a great engine in him, his curiosity about the history of farming and farm implements, and attached him to farmers. One piece of farm equipment almost begs for another, and building up the tools to grow and handle grains corralled a lot of resources, in terms of money, repairs, or time. On the one hand, this was very satisfying, and on the other it was kind of addictive.

"I kind of had this love affair with iron and steel. I just loved the machinery part of growing grains," Jack said. Even in the thick of his health issues, he bought a corn sheller that removes corn more gently from a cob than a combine. This is better for saving corn seed and keeping the kernels from spoiling.

The need for machinery he describes as a bug, but I've also heard him describe farming as a bug that people catch. This stuff grabs you and asks that you do it. Now he's captivated by soil, as he told people at the NGGA conference in March 2014:

"If you really want to build soil and put carbon back into the earth, you need to have your land in forages two to three years of the total six to seven in a rotation. Things have to go back to sod to heal," he said. "Most of us have limited acreage to work with, and the temptation is to plow it up and plant it again next year."

His keynote talk was an encapsulation of his farm life, and an abridged version of the book, capped with an emphasis on soil health. The farm is a model for organic practices, and the book outlines cropping systems that are very good at soil building, but since his recovery Jack is feeling very strongly about growing the land with grasses and legumes, rather than just growing food.

"Having cows on the land harvesting their own forage is probably more sustainable than doing a lot of tillage and a lot of plowing," he said. He'd just started planting oats in the fall and letting them die in place as a kind of permanent mulch. He wasn't sure how many acres he would plant to grains that season, even though he could sell just about anything he put in the ground.

"People are banging down my door for corn for distilleries, cornmeal, dry beans, black beans—you name it! The opportunities are there for everyone," he said. "I can't do it but I'll teach you how."

He was almost contradictorily evangelistic, cautioning people against the allure of an equipment- and land-hungry crop, and at the same time telling them to seize the market. He saw the future in small-scale growers like Noah Kellerman, who is making his own way with grains and vegetables, building a dryer, growing barley for malting, and making flour and pizza. Diversity and self-sufficiency were his prescriptions.

"I've built more soil as a grass farmer than as a grain farmer," he concluded, inviting people to keep a better eye on the bottom line of soil. "I'm part of a rather large network, and if you've got questions and problems please get in touch with me. I'm there to help."

The picture he ended on, pointedly, was Noah pulling an All-Crop.

That invitation for people to find him was sincere. He's slowing down but by no means shutting down. He's such a social person that he needs plenty of human contact to keep his soul well nourished. Shifting gears to his role as a mentor, he taught a class at the University of Vermont; when he brought that class to Loic's farm, Randy George tagged along.

Teaching, writing, and farm tours are show-and-tells: conduits for information in that large network Jack mentioned. Other parts of that network are the NGGA's field days, and its one-day conference in March. One year, Loic Dewavrin spoke. The next, Judith Jones from the Whole Grain Council discussed myths circulating about gluten, giving grain farmers and bakers some answers to the many questions and fears floating around on the subject.

The conference puts information and experts in front of a good-sized crowd. Up to two hundred people attend the event; field days attract more modest numbers, anywhere from twenty to seventy-five people. The NGGA, however, began as a more one-on-one exchange between peers.

In 2004, a small group of farmers growing grains, including Ben Gleason and Jack Lazor, saw the need to get together and share information. They had a ready partner in Heather Darby, the University of Vermont's new agronomist.

"There was such a booming demand for both feed- and food-grade grains," Heather said, describing the group's beginnings.

The NGGA grew family-style, around shared work and problem solving. The problem was the lack of organic grain production created by a rising market for organic milk.

"There was just not a lot out there," Heather said, in terms of education and information for growers. The know-how was gone, vacuumed out by better grain-growing lands in the Midwest. Shortly after the NGGA started, the locavore movement took off, and a lot of consumers were searching for food-grade products like oats and flour. "We didn't shift away from feed grains, but realized how complementary it was to be able to have a market for both."

The NGGA met monthly, really digging into the topic. Over time, more non-farmers joined, like baker Jeffrey Hamelman. He runs the bakery at King Arthur Flour's headquarters in Norwich, Vermont, and has written *Bread: A Baker's Book of Techniques and Recipes*, a book so thorough that some say it is the only one a baker needs. He joined the NGGA because he wanted to lend a baker's perspective to the group.

The baking contingent has grown stronger over the years, adding their concerns to the project. King Arthur and Red Hen are involved in bake-testing, and the explorations are not limited to the NGGA or Vermont. Jeffrey and Randy developed protocols for baking and analyzing performance, along with bakers from Maine, Jim Amaral and Alison Pray. The Northern New England Bread Wheat Project, a grant funded by the USDA's Organic Research and Education Initiative and led by Heather Darby and Ellen Mallory from the University of Maine, facilitated this analysis.

The initial question the farmers asked—how can we grow higher-yielding and higher-quality grains in Vermont—drew many people into the inquiry and spiraled out into many projects. That USDA grant sent a group from New England to Denmark, to study organic grain farming and marketing habits in a similar climatic zone. Another group of farmers, bakers, millers, and researchers went to La Milanaise, the larger regional flour mill in Quebec. These trips, and the bakers' involvement, made the need for cereal testing apparent. While bakers rely on their senses to see how flour works, qualifying the characteristics of different wheats is important to understanding how dough performance relates to farming practices and the quirks of variety and season.

UVM Extension now has a cereal grain testing laboratory that can evaluate grain for baking qualities like falling number, protein, and fusarium contamination through DON levels. In the booming market for malting barleys, the lab evaluates grains to see if they meet basic quality parameters, too. The testing performed here fills a valuable niche in reestablishing grain economies.

The lab is an example of the responsiveness of Vermont's extension program. Though small, that limit allows for a lot of creativity. Larger states might have a corn extension specialist and a soybean extension specialist, but Heather covers all field crops. The downside of that is being pulled in too many directions.

"The upside is we get to create something very different," said Heather.

There is more freedom to develop programs that span topic areas that would otherwise be discrete. This means investigating a diversity of crops, like flax and hops, and being able to meet more of a farm's needs head-on, in line with that original purpose of extension. This is a bridge between farming and science, in a landscape so small that both farmers and researchers can stand together and eye the same goals.

Such cooperation wasn't always the case. When Heather first arrived, she asked a group of farmers what they wanted and got no reply. Farmers were not used to being asked for their involvement. Once farmers realized that an unbiased partnership was forming, the opportunities for collaboration took off.

"We've really worked with any farmer group, not just the NGGA, but the Farmers Watershed Alliance, as well, to really figure out what the true needs of the farmers are and put those on the ground, in both research and education," said Heather. These concentrated efforts lead more quickly to changes that impact and strengthen farming.

Heather has mountains of energy, drawing many people into the fun of her work. Erica Cummings left another soil lab at UVM to join her and help direct an octopus of programs and projects. She set up the lab and helps guide eager students both there and in the field. Trials take place all over the state, and sometimes out of state, too. These efforts fall under the title of a group, The Northwest Crops and Soils Team, that did not exist when Heather started working at UVM in 2003.

Over her tenure at extension, Heather built a research and education program, and the team is the result of her vision. She wants to bring the best to local farmers, and she's done that by pooling resources through seeking and acquiring grants. The growing number of grants and workload necessitated drawing in more people. Now the team has seventeen employees, some of them seasonal and students, to help make sure the work is done and, most important, delivered to the farming community.

Extension fits well into Vermont's overall support for farming. A strong vein of New England's famous self-reliant character is being mined to re-regionalize food production and distribution. Farm advocacy is powerful at all social strata, from grassroots to government levels. Old-school back-to-the-landers are cooperating with new generations who want to connect to the land, forging networks that support and create small farm and food enterprises. The state made an early commitment to local food purchasing that set a tone for other states looking to redevelop regional agriculture.

Model grain farmers and millers like Butterworks and Gleason's Grains, which has been growing and milling wheat for more than three decades, are inspiring new farms and mills. Rogers Farmstead had a great first year with wheat and oats, and is selling grains in small bags at the farmers market in Montpelier. Green Mountain Flour has been baking and milling for a few years in Windsor, working hard to build relationships with farmers and get local staples into the kitchens around them. Its pancake mix is making its way into inns and schools. While there is not enough wheat grown in the state to supply even one bakery, beyond the smallest scale, with flour, many bakeries feature local loaves made with cornmeal or wheat flours from Gleason's Grains and Nitty Gritty Grain Company. (The wheat for the latter is from Aurora Farm in Charlotte, and milled by Champlain Valley Milling in New York.)

Randy George has been an active flour hound, regional and otherwise. Red Hen Baking Company makes 15,000 pounds of bread a week. Some of it goes to restaurants, and some of it sells retail in the bakery's own café and in stores. At this scale, and in Vermont's somewhat rarefied food environment, they can make choices about sourcing that other economies would not support. This doesn't mean every ingredient is the most—most sustainable, most handmade, most lovingly manufactured. But these elements are factors the bakery can weigh more than bigger food producers, and more than operations in places that have fewer well-informed consumers, and fewer well-farmed and carefully crafted selections. It's a recognition that food doesn't have to just come from a box, but that it comes from someone, from somewhere.

The human element is what Randy loves about local ingredients. To be able to connect with the person who grows your cucumbers or cheese is great. To be able to connect with the people who grow your flour is better than great, because it is such a novelty and an education.

"I was always interested in trying to get to the source," said Randy, who began baking early, helping his mom as she used the famous *Tassajara Bread Book*. Learning to bake professionally, through jobs in Vermont, Maine, and the Pacific Northwest, he wanted to know flour and where it came from, but learning the craft commanded his attention. When he and his wife opened Red Hen, he started buying from mills like Lindley in North Carolina and Heartland in Kansas, relatively large small-scale mills that could trace their grains to growers. He's built a good friendship with baker Thom Leonard, who has been involved with Heartland Mills from the start. Long distance, Randy followed the revival of Turkey Red wheat production, and got familiar with Turkey Red and other Heartland flours by using them.

Being involved with grains closer to home has been equally engaging. His interest in using sifted flour, and his track record as a steady customer, is largely responsible for Ben Gleason expanding his milling operations to add a sifter. Gleason's Grains began milling its own wheat in 1982, and Ben's inclination was to stick with whole-grain flours. The business friendship with Randy convinced Ben to invest heavily in infrastructure, constructing an addition to the mill, and doubling his sales to Red Hen.

As Moulin des Cedres increased production, Red Hen became its first sizable client. Loic's plan to partner with just a few bakeries, and Randy's interest in regional flour and farming are well matched. By the end of 2015, all the organic, unbleached flour the bakery needs will come from this farm and mill in Quebec, and all the flour used in the bakery will come from within 150 miles.

Beyond any romance attached to the term *local*, these sources represent stability. Early in 2014, bakers faced shortages in the North American organic wheat supply because of drought in the Plains and other problems. Some mills were turning to Argentina for organic wheat. Knowing who grows the wheat and makes the flour gives the bakery some predictability. The fact that the flour comes from one farm is a luxury for Randy and all the bakers at Red Hen, offering a steadiness of performance they all welcome. There are other advantages, too.

"I love learning about the crop and the people who grow it, and the challenges involved in getting it from the field all the way to the mixing bowl," he said. "It's a real art."

Art indeed. Randy made me think of the term *transformation*, and the implications of craft wrapped into the word by farmers in France and Canada. In considering non-commodity food chains, I get stuck on the idea of visibility, and the necessity of people seeing each other's work. Yet transparency has to be accompanied by skill, or else local might just mean lousy. Craft farming, and milling and baking, is critical to the revival of local wheat. One particular day, these elements really came together.

In the summer of 2013, I went to a field day at The Borderview farm. UVM hosts a lot of trials on this former dairy, which sits on a dirt road dividing American and Canadian cornfields. After touring wheat plots, the hop yard, and flaxseed trials, we sampled bread. Farmers Ben Gleason and Jack Lazor stood for snapshots with Randy, holding up tawny slices of sourdough loaves. Ben had milled the flour that he and Jack had grown out on their farms.

The event was an event because there was finally enough wheat, grown from historic Vermont varieties, to make some test batches. The genesis of these loaves occurred, most recently, in 2006, when varieties developed by Vermont seed breeder and Civil War resistor Cyrus Pringle were procured from the USDA gene bank. From one hundred seeds of a nineteenth-century wheat named Defiance came 20 pounds of flour. The resulting bread held people in its thrall.

Randy also bake-tested heirloom wheat varieties such as Lagoda, Marquis, and Mida, and said that all the varieties baked pretty well. Each loaf was very tasty, exhibiting good crust, crumb, and holes, those telltale traits of hearth-baked artisan loaves. The bread made from Pringle's Defiance got the heartiest applause, but who could tell if it tasted the best? The nuances of wheat are subtle, and we are just learning how to find and name them.

What mattered were the labors that people put into researching and growing, milling and baking, and the link to another generation of Vermonters dedicated to farming. An old wheat, developed at a time when grain production had already marched west, had a place on contemporary farms and in modern kitchens. This glance back and leap forward was a banner moment for all the people working to transform wheat, from seed to flour, from flour to bread, from invisible ingredient to patriotic crop and care-filled food. The farmers, the miller, and the baker broke bread along with a small appreciative crowd. It was both symbol and substance. Transformation.

CHAPTER NINE

THE BREAD LAB

"*J* pulled off the freeway and saw wheat," said wheat breeder Steve Jones. "I took a picture of it, drove another hundred yards, took another picture. More wheat."

A wheat breeder snapping pictures of wheat might not seem unusual, but Steve was interviewing for a job on the side of Washington State that grew liberals and vegetables, he thought, not wheat.

"I thought I'd work on sauerkraut cabbage or something like that," he said. From 1995 to 2008, he was the winter wheat breeder for Washington State University in Pullman, east of the Cascade Mountains. This is wheat country, and he served farmers who grew large acreages for the commodity market, 90 percent of which was exported to Asia. The state is the fourth-largest wheat producer in the nation, growing 2.2 million acres, much of it soft white wheat. He liked his job, but the work environment had grown unpleasant.

"I loved breeding wheat and I loved working with farmers, but I wouldn't breed any wheat that we had to sell to the farmer," he said.

This attitude created friction, and so did his interests. Receiving a grant for organic wheat breeding had prompted a punitive visit from the Wheat Commission, the industry body that serves wheat growers in Oregon and Washington State. When the director's position came up at the Mount Vernon Research and Extension Station, Steve applied. He thought he'd be leaving a tense situation. He thought he'd be leaving grains, too. Though he was already working with organic farmer Nash Huber on the Olympic

Peninsula, in those troublesome organic wheat trials Steve knew that grains were not the agricultural driver west of the Cascades.

Soon, though, he realized that growers in the Skagit Valley needed research for grains, which they grew in rotation with high-value crops like tulips and potatoes.

Steve told me this story in The Bread Lab, leaning against a counter, and bending his long body a little, as if to shrink his height to match his down-to-earth manner. Behind him on a chalkboard someone had written a list of flavor notes: nuttiness, bran, bitter, sweet, fresh-bright, spice, grassy, hay, sour, straw.

In 2013, Steve Jones still seemed surprised by the good luck he'd found. I fell right into his story, caught up in the near smile that was always on the verge of breaking across his face, and by his voice, which has a similar happy cliff.

The Skagit Valley and surrounding counties are part of the Salish Coast, a mild climate tempered by the proximity of the Pacific and the border of the Cascade Mountain range. The region runs from British Columbia to the southern end of Oregon, and never gets very hot or very cold. Around Mount Vernon, the Skagit River delta really mollifies the climate; farmers grow eighty different crops, from spinach and broccoli seed to potatoes and tulips.

These are high-value crops that need grains in rotation to help fight pests and disease buildup in the soil, and add organic matter. Farmers had to grow grains, but they didn't profit from them, and even lost money in the years they grew them, selling them into the commodity export market. Steve's expertise might be able to change that.

More than five years into his tenure, the changes are starting to show. Fairhaven Organic Flour Mill now buys more than 60 percent of its grains in-state, up from just 10 percent a few years ago. Feed mills are using region-ally grown grains rather than bringing grains from the Midwest on railcars. Research at the station is stretching into specialty markets, developing a malting barley industry, bread wheats, and high-nutrition feed grains. The goal is to give growers resources and reasons to keep the value of their products in the region, creating a closed-loop grain system.

"We're just not comfortable with exporting the value in these things anymore," Steve said. "It's not just the grain biology, it's the grain economy, and that's how people view it."

His engagement with re-localizing grain production has gained him many partners in the region and around the country. What the Mount

Vernon station is doing matters to local growers, and to bakers and millers, and brewers and distillers. Chefs from far away make regular treks to The Bread Lab, and Steve travels to restaurants and land grant universities, speaking about the concept and practice of growing grains outside the grain belts. This place has become ground zero for the revival of regional grain production.

The resources favoring this development are considerable. Other places don't have farmers who already grow grains. These farmers are also familiar with making high-quality products, like vegetable seeds and flower bulbs. The Skagit Valley and the surrounding counties have a good climate and soils. All of these elements had the benefit of an economic engine called the Port of Skagit. Shortly before Steve arrived, this economic development agency identified farming as an area to aim its efforts, stating a new mission to support value-added agriculture.

Markets from Portland to Vancouver are full of informed consumers, ready to receive innovative foods, but there's a history of the Skagit Valley growing grains for the region, too. As pioneers moved to the area, the Skagit grew oats, the diesel fuel of its day. Draft horses did all the work in the Northwest timber industry and helped build cities and towns. A hundred years ago, a world record for wheat yields was set on Whidbey Island 120 bushels per acre. For comparison, Kansas currently averages 33 bushels of wheat per acre, grown with heavy chemical inputs.

Modern yields in western Washington are not as dramatic, but the potential for change was and is great. Growers have combines and crop knowledge. Fifteen thousand acres of wheat was once harvested and traveled by truck and railcar to Portland and overseas, selling at whatever rate was set by the commodity grain exchanges in Chicago or Minneapolis. The wheat breeder had a captive audience.

"The first week I was here I had a breakfast with the farmers," Steve recalled. "One of them said, 'I want everything you do to stay in the public realm, and I want it to be for the public good. I don't want you to commercialize anything.' Holy shit! I hadn't heard that in twenty years."

Nationally that's a pretty novel relationship between a land grant university and farmers. Though agricultural researchers at such institutions are public servants, the work done tends to benefit corporate funders. But lots of things about this agriculture station are surprising.

The farmers of the area have a unique relationship to the ag station, having gone to the state and requested their own research facility in the 1940s. Research conducted east of the mountains didn't have relevance to their farming needs. The university agreed to provide staffing, and the farmers gave land and money to construct a building.

This beginning is very unusual. Most stations are much older, and tied more closely to land grant institutions. Their funding parameters and the scope of their investigations are limited by working within a larger academic frame, which frequently means tethering projects to agribusiness support. The Mount Vernon station is still part of the state university system, but it is a bit of an island.

Farmers have directly invested in this facility. Area growers built the original structure in 1947, and the sons and daughters of those growers raised $2 million locally for a new $8 million facility that opened in 2006. This sets the stage for a very cooperative relationship, and embeds the faculty in the community.

Asking for a resource and contributing to its development influenced the connections between research and farmers, but these farmers already had very cooperative arrangements. While competitive in their businesses, they've had to help one another with land. The realities of producing spinach seed, for example, show just how much they have had to work together. Once every fifteen years you can grow a crop of spinach seed. Without swapping land, you'd invite diseases, the quality of your seed would plummet, and you'd go out of business very quickly.

Working in specialty crops, these farms have no safety net of government price supports like crop insurance. They are one another's safety net, and their own, so they see the value of diversifying income streams through multiple crops. This is mostly conventional farming, but there's less mono-cropping than on many American farms.

Aside from grain production, the farms are not tied to commodity markets. For instance, potato farmers in Maine sell their crops to potato processors and often take low prices for their harvests. In contrast, most potato farmers in western Washington are vertically integrated, and process and market their own crops. Most have trucks to cover distribution. In other words, the growers are ready to participate in value-added grains projects because they are already adding and capturing value in their other crops.

The agriculture in the region is very complex, a puzzle of crops and soil and people. When a new herbicide is evaluated in the Midwest, scientists have to see how it does on two crops, corn and soy. Here, if a new potato herbicide comes out, farmers need to know how it will impact the other seventy-eight crops they grow.

Twelve family farms generate $70 million of potatoes. Berries are a big crop, and so is seed production. Half the cabbage seed in the world is grown within 10 miles of the ag station, as is half the spinach seed grown in the United States. Fresh-cut broccoli and cauliflower figure into the portrait, as do other vegetables. Tulips are a big deal. Streams of cars flood the roads, bringing a million people to tour the colorful fields each April.

This was all news to me, all except for the cars. Though I lived in Seattle during the 1990s, the one time I'd tried to see the flowers I got discouraged by the traffic and went hiking instead. My other contact with the area was back when I was eighteen and chasing authors, not grains. Novelist Tom Robbins lived there, and on a cross-country trip, ostensibly to investigate colleges, I tried to find him.

I was not a skillful researcher and just landed myself at the post office, where I'd already sent altogether too many letters. I couldn't push the clerk to reveal any details about his location. Not that I would have been savvy enough to get there. Maps are pretty mysterious to me, so I just had a cup of coffee and headed back to Seattle.

I'd taken the Greyhound bus from upstate New York, but within the state, I was hitchhiking. The main character in Tom Robbins's book *Even Cowgirls Get the Blues* has huge thumbs. Mine felt inadequate, even though I knew she was a dreamed-up girl and I was just a dreamer.

A tan Buick pulled over. Women didn't usually stop for me, and this one surprised me even more because she had white hair. She brought me to her house and gave me a meal of ham, pineapple rings, and homemade dinner rolls. She gave me a warning about the Green River Killer, who was active at the time, and drove me to the bus station, where she bought me a ticket. I had no idea that the older me would be back with a rental car, prowling for grains.

"These aren't wheat farmers, these are farmers that grow wheat," said Steve on my first visit, introducing me, in a general sense, to the people he worked for. The land is too valuable for the main crop to be wheat, whether the farm is 2 acres of vegetables or 2,000 acres of tulips.

"I found out there was a true need for wheat research," he said, that smile breaking on his face.

Steve led me around The Bread Lab, which he'd begun to outfit with equipment to interpret grains and flour. The lab had tools familiar to commercial baking, like a proofing box draped in plastic curtains, a mixer, and a deck oven. The lab also had unfamiliar tools to help define the baking characteristics of flour. My favorite was the alveograph, which blows a dough patty into a bubble, measuring gluten strength.

One machine tests falling number by counting how fast a plunger falls through a test tube filled with a slurry of flour and water; the number shows a flour's enzyme activity, a vital tool for bakers wanting to know how flour will behave in dough. Another lab room had mills and more ovens. Soon, Steve said, he could hire a full-time baker to analyze the grains and flour, once varieties and crosses hit certain agronomic markers, first in the greenhouses, then in the field.

The station tests thousands of types of wheat each year, and also researches seeds for many other crops. We walked to the greenhouses, where wheat plants stood in black plastic nursery pots. The lighting was harsh and bright. Breeding sounds sexy, but the reality is pretty static. The pots sat on metal mesh tables, and the stuffy warmth, even as a reprieve from the damp Pacific Northwest winter, felt mildly unpleasant. In one row, singular wheat heads poked up from the long-armed leaves, some browning, and some green. Other pots in other rows had heads enclosed in plastic.

"You remember Mendel, right?" he asked. I nodded. Gregor Mendel, peas, hybrids, genetics. Check. He showed me a sleeve slipped over a pair of green wheat heads dusty with pollen. "We combine traits we want to put together, and put a male and a female plant in a dialysis sleeve. "

After pollination, the plants mature, and the seeds of the cross are harvested and planted into pots for another round of selection. That row of pots with single grain heads was first-generation plants; only some of these would pass through this round, and offer a next generation that would also grow indoors. The third-generation picks would go outdoors to the test plots. In any given year, 40,000 wheats could grow here, indoors and out. Eventually, samples from chosen varieties would go to The Bread Lab.

"The strategy is to do the testing for flavor, nutrition, and functionality," said Steve. "We do this by coming from field to lab, milling the flour and testing on the alveograph and farinograph, and doing full bakes as well."

All of the plants looked similar, yet I knew they might grow miserably or terrifically, and taste either way, or just so-so. Wheaty and indistinct. I knew that fresh flour tasted different from bagged stuff from the store, but what about that list on the chalkboard? Could people really tell the difference between grassy, hay, and straw? Yes, Steve said, and the grassiness in breads made from Mount Vernon wheats differs from the same varieties grown elsewhere, like in Oregon's Willamette Valley.

The assessment process for wheats and barleys in Mount Vernon, as in any breeding program, begins with agronomics. The plants have to grow well for farmers. For starts, that means disease resistance and good yields. Besides good farming characteristics, wheat and barley breeders usually seek to meet industry standards for milling, malting, or animal feed. Breeders know what they *can* get from plants, but given the constraints of market demands, they rarely scratch the surface and see what they *might* get from a plant.

Traditional wheat breeding analyzes grains to hit the specifications of large mills, but the aim at Mount Vernon is to help farmers get out of the volume business and serve smaller clients. Rather than just meeting those narrow performance targets, researchers are seeking varieties that exhibit novel flavors and traits. The quest is more nuanced.

The search is on for hulless barleys that can deliver whole-grain nutrition, since barleys are usually stripped of their bran by the pearling process. Malting barley suited to the climate and soils will mean getting $2,000 a ton instead of $400. The price jump is not as significant for bread wheats, but the leaps in taste can be theatrical. George DePasquale, from Essential Baking in Seattle, said that one variety had the best taste he'd found in thirty-five years of baking.

Bringing bakers into the equation is novel in grains. Bake-testing usually happens at flour companies, or in quality labs at universities. That's because the standards are predetermined, and new varieties need to be evaluated in comparison with those standards. Bakers meet flours by numbers on specification sheets, but in emerging grain systems, bakers are meeting actual grains, not just numbers. Those introductions are not just handshakes, but the result of queries that take years to ask and answer. The Bread Lab, as it matures, is becoming a dating service for farmers and seeds, and all the people who use grains, from chefs and bakers to all the grain-based fermentation crews, including maltsters, brewers, and distillers. Steve has a team of grad students and staff coordinating potential mates.

The work is wide and inquisitive. That list of flavor notes is barely a start to qualifying what grains do. Taking varieties, they see how they can explain it within the worlds of functionality and flavor.

"How can we take that and manipulate it?" Steve said. "As chefs and bakers and millers, what can we pull from new varieties? What's available to us?"

Most seed breeders target for basics like disease resistance or plant height. The target for all of the wheat bred in America is roller milling, so breeders select for a beefy endosperm and gluten properties that lead to loaf volume and strength for mixing at high speeds.

"If you're a baker, you're getting flour that was bred for that, even if it was bred a hundred years ago," Steve said. The capacity to add to those categories and breed for technologies like stone milling and artisan baking is giant. New targets for grains are just beginning to be known.

Graduate students work on their own projects and collaborate on the general analysis of grains. The full-time baker arrived in 2013. Jonathan Bethony wears a beard and glasses, and has the sparky kind of mind that is great for chasing mysteries. He baked his way into this job, working in bakeries and training at the San Francisco Baking Institute. He won his spot at The Bread Lab over numerous bakers who vied for it. He is very well versed in the science of bread, and explains flour and sourdough with botanical and chemical detail, and beautiful metaphors.

At The Grain Gathering, an annual conference held in Mount Vernon, Jonathan gave a class on pre-ferments, levains, and sourdough starters. He described sourdough in social terms, comparing its care to that of any human relation. Sure, a starter can rest in the fridge, but it will need some tending when you want to make bread.

"Let the sourdough sit, and just like a friendship, the longer you spend apart, the more you're going to have to feed it," he said. This was a metaphor, but real, and made the other information he presented easier to understand.

"Sourdough is flour, water, and air. One gram of flour has 13,000 yeast cells," he said. Yeasts feed on the starches in flour, forming sourdough cultures. Whole-grain flours have more minerals and more microbiological activity because of the yeasts that hang around the bran. Rye has more activity than wheat.

"Sourdough cultures have two major components, wild yeast and bacteria, or lactobacilli. Yeasts tend to colonize first," he said. These competing forces create lactic and acetic acids. Certain sourdough flavors come from lactic acid and warm temperatures; colder temperatures favor acetic flavors and acidity. The microbiological population shifts over time from yeasts to bacteria, making a more sour starter.

He gave a schedule for starting a starter, and kept riffing on the relationship theme as he suggested time lines for working with sourdoughs and troubleshot questions from the audience. Even though I'm less interested in bread than pancakes, to say the least, and rather devoted to baking powder because of the way it leaps my cakes off the griddle, he had me riveted. He ended his talk by playing the theme from the TV show *The Love Boat.*

I solicited a private pancake chat to try to see if he could help me understand why malt gave such a boost to the whole-grain pancake mixes I make.

"As soon as you add water and oxygen to the mix, enzymes start reacting," he said of flour. "That's why a seed grows."

As a kernel starts to germinate, water activates the enzymes in the kernel and starts breaking down the components of a seed. The proteases break down storage protein molecules (gluten proteins). The amylase enzyme breaks down starches into sugars. When you're baking, you're not trying to grow a plant, but all of these things are happening.

"The enzymes unlock the food supply, which would normally be utilized by the plant as it grows. Those enzymes survive the milling process and enable similar processes to happen in baking," he said.

Since I use whole-grain flour and freshly milled malt, maybe the extra enzymes were responsible for the liveliness I observed. Bran contains a lot of enzymes. I've been baking for more than forty years, but this was the first time I'd seen the parallel between seed growth and baking.

Jonathan was just doing his job, interpreting flour for bakers. He's analyzed baguettes from Paris, trying to duplicate a recipe for a client who wants to start a bakery. He's helped junior high students explore how fermentation degrades gluten over time.

Bread baker Scott Mangold, who owns Breadfarm in nearby Edison, said that Jonathan conveyed an attitude that had eased his nervousness about using whole wheat and local flours. Scott worried that the dough would fall apart, and Jonathan said he'd felt that worry, too, and at one point he just started to look for signs he knew from baking, and chose to trust that the flour would work.

Jonathan is a bit of a flour whisperer, but he is not alone in his field. Cereal scientists tackle the task of reading flour, and so do experts on call lines for King Arthur Flour. All around the world, bakers educate themselves on their ingredients, and help one another. But The Bread Lab is a new kind of place for these intersections, and it's about to get newer.

A 12,000-square-foot facility at the Port of Skagit will stretch the lab from 800 to 2,400 square feet, and extend the possibilities for collaboration even more. There will be offices and meeting rooms separate from the ag station, which does have work to do that is not related to grains. The new facility will have a teaching center five times as big as the space currently used for classes. The port is giving the first year of rent free, and helping with renovation. This is a continuation of the support they've extended to the grains work, which has been funding research and helping develop enterprises, like Skagit Valley Malting.

The Port of Skagit economic development agency formed in 1964. Over the years, the port has moved from directly supporting jobs in particular fields, such as the timber industry, to more systemic approaches. First that meant a shift to include environmental sustainability; more recently, it's encompassed farm stability.

"Our mission is good jobs for the Skagit Valley and we work in several sectors, including value-added agriculture," said Patsy Botsford-Martin, the agency's executive director. The farm focus came from a survey of stakeholders in the mid-2000s. A lot of people said they wouldn't be in the area if it weren't for the farmland and open space. Serendipitously, the port identified this goal shortly before Steve Jones came to Mount Vernon, and the value-added investigations and collaborations took a grains angle.

The first was partnering with Wayne Carpenter, founder of Skagit Valley Malting. The port renovated space for use as a malthouse. The next project the port undertook was building a live grain storage facility that would serve growers and their customers, like bakers who want to mill and maltsters. The existing grain-handling facilities in the Valley dry grains at a high temperature that doesn't keep the germ alive. A few companies have seed-handling facilities, and at first the port thought they might be willing partners for a live grain enterprise, but the harvest times are too similar to share equipment.

The port invested $500,000 in equipment for the new grain facility; Skagit Valley Malting designed and installed the equipment. The building has two 200-ton dryers that gently handle and gently dry grain, and room for storage. The port is anticipating a need for more storage as area growers begin to switch to specialty products, and is seeking funds from legislators or the county to help develop a facility to store 2,500 tons of grain. The port builds success by nudges like this, using its own investments to try to leverage further support. Dried beans are the next area that might get infrastructure of this sort, for cleaning and storage.

"We're working on developing a buckwheat mill," Patsy said, for soba noodle flour. The inspiration is a stone mill in Japan that has been in operation since before World War II. The port is seeking a private-sector partner for this effort.

The value-added work also includes non-grain enterprises. Recently, a pickle and sauerkraut company opened up in a Port of Skagit building, using cucumbers and cabbages from local farms.

Patsy grew up in the area and studied geology, political science, and the environment, digging in to the sense of connection she felt to the land.

"I was always interested in sustainability in the valley," she said. When she started to work at the port in 1995, she could explore that interest. "How do we keep and protect what we hold so dear in a market-based world?"

Picking strawberries as a kid, she could not have imagined that farming would take the twist that it has, finding sophisticated ways to grow and use food. But here she is, brokering partnerships between businesses and the land.

Skagit Valley Malting is a high-tech venture, the opposite of malting start-ups in the Northeast, which often convert old dairy tanks to get into the game. The company is the second and third career for some mostly retired high achievers, such as pilots and veterans of software businesses.

Wayne Carpenter and his partners have designed systems that can regulate and control the malting process with great precision. The initial plan was to make a dozen machines that could each handle three tons of malt, but in the process of analyzing how their product would fit the market, they doubled capacity. That's because, while there are more than five hundred craft breweries in the Pacific Northwest, there are practically no craft malthouses. Only Rogue Brewing has its own barley farm and malting facility in Oregon, making the malt used in its beers.

Skagit Valley Malting refined their plans for years, running trial batches on two prototype machines, one that makes 500 pounds of malt and one that makes 50-pound batches. These lab-scaled models help the maltsters test and perfect programming for the full-sized machine, a 6-ton system that came online in late 2014. Brewers and distillers loved working with the malts from trial batches, which got really good results in third-party testing, hitting admirable scores for friability, beta-glucan, and enzymes.

When people see these figures, "They go from potentially skeptical to telling me, 'I should be your first customer,'" said Wayne Carpenter.

The malthouse at the port will fill with more of the full-sized machines as the pieces of the malt puzzle come together, including an increase on the supply side of quality malting barleys bred at Mount Vernon, in concert with barley breeder Pat Hayes, from Oregon State University (OSU) in Corvallis.

Farmer Kraig Knutzen has been working with the maltsters. Knutzen Farms is more than a hundred years old, run now by Kraig and his siblings, and a few sixth-generation family members, too. The potatoes they grow have a good reputation for quality, with some of them traveling a good distance to the East Coast and the Pacific Rim.

"When I came into the business it was more of a regional basis, but the day of being a small truck farmer is more and more difficult," said Kraig. His farm demonstrates the principle of high-value producers who don't get much value from grains. "We can only produce potatoes one out of every three years. It's good that we have the ability to increase returns on rotational crops."

Grains add biomass to the soil, but not much income. Knutzen Farms plants between 1,800 and 2,000 acres a year, and anywhere between 350 and 700 of those will be grains. Wheat is soft white sold into the export market, and barley is sold locally for the feed market. They've produced some wheat organically and sold it to Fairhaven Flour Mill, but Kraig Knutzen has his eyes on malting barley.

The farm is an active participant in the barley crop's research-and-development stage. They've been growing out a few types identified by OSU and WSU, like Alba and Fullpint. They also trial new varieties on-farm, and grow out interesting varieties on a larger but still trial scale, 4 to 5 acres.

Researchers are looking for qualities like the kernel plumpness that maltsters want, and qualities that farmers need, like winter dormancy and good sprouting control.

"We're working on the amount of nitrogen we put down to make sure we don't get too high a protein," said Kraig.

Researchers and farmers are talking about changing planting rates. This affects tillering, which determines how many grain heads a plant will produce. Another issue is making the stalks sturdier so they can stand up better in heavy rains. The usual breeding targets are also pursued, like yield and disease resistance.

The cooperation is great, Kraig said, but not entirely a new thing. Grains are exciting, but they are just an extension of the attention that scientists at Mount Vernon pay to all crops. Currently, WSU researcher Dr. Debra Ann Inglis is making headway on potato blight, and those high-value seed operations are like machines that need continual servicing from the "farm mechanics"—seed breeders and agronomists who work to keep agriculture working in the region.

The transition to value-added grain markets is slow but steady—another change these farmers are embracing. When commodity grain prices went sky-high in 2008 and 2009, for instance, the local feed mills knew they had to get out of the habit of buying grain from Nebraska. The focus on grains at the ag station has helped redirect grain sales that used to go for export into the feed market. Farming is always evolving. Peas, for example, were once a big industry, and now they are not. These farmers are always ready to try new solutions and the next new crop. They aren't as locked into setups as farms that operate more at the whims of commodity markets. They can stay nimble on changing ground.

Hedlin Farms, a third- and fourth-generation family farm, has incorporated organic vegetable crops into conventional seed production.

"Farming organically has made us better conventional farmers, but we never lost track of soil tending on our conventional farming," said Dave Hedlin. "Skagit Valley has the best soils in the world. As a grower I want options, and research is a big part of that. We really count on that to survive."

The family farms 500 acres total, half of it organically and half conventionally. In any given year 150 to 180 acres will be in grains.

"We farm a heavy clay alluvial soil," Dave said, "so we need that grain rotation to put organic matter and straw back into the soil."

Hedlin Farms direct-markets organic vegetables through a farm stand and at Seattle farmers markets. They sell to restaurants, too. The grains fit into this side of their business as flours, milled by Fairhaven. Bags sell at markets, and are tucked into CSA boxes for special occasions, like Thanksgiving. The malting business is emerging as a serious option, and the Hedlins are selling barley and wheat for both malting and distilling.

Dave doesn't see leaving the commodity market entirely, but he's pleased with the new angles that the station is exploring, like heritage grains, to help differentiate crops in the marketplace. Since the farm plants winter and spring barleys, feed mills, both organic and conventional, are good backup markets for the grains that don't make malting grade.

Wheat and barley are not the only grains that shine under the sun of WSU. Graduate student Louisa Winkler is studying oats.

"Oats have a high moisture requirement compared to other grains and can do well in high-rainfall areas. Western Washington has similar climate characteristics to the UK and Ireland," said Louisa, who has worked with oats in that part of the world. "There's a lot of literature in the United Kingdom on how oats, and hulless oats in particular, are useful as a livestock feed."

Louisa is studying oats from a few angles, one of them to see how they can perform in poultry rations. Today, America's chickens are normally fed corn and wheat, but oats are an alternative that may have nutritional advantages, and they are also easy to grow in western Washington. Oats were often the grain feed of choice for previous generations raising livestock in the region. They are high in beneficial oils and protein, and may be able to provide more of the essential amino acids that livestock need but that corn and wheat lack.

Louisa's poultry feed research is focused on hulless oats. Some oat varieties have a thick hull that does not detach at harvest. These are hulled oats, and they require special machinery to remove the hull but still maintain good storage ability; these are the preferred type of oats for human food. Hulless oats lack this hard outer covering and can be fed directly to animals, since there is less concern in this market about superficial damage to the grain. Depending on the variety, hulless oats may be very high in oil and hence in calories, which is good for animal feed.

One of the main problems with hulless oats, though, is that they are less widely used than hulled oats, so fewer varieties are available and less is known about their qualities. Few oat breeders in North America work with hulless oats. After a lot of time and effort, Louisa has managed to import seed of hulless varieties bred in the UK; this is a difficult process, due to the complications surrounding cross-border movement of plant materials.

The poultry research will have several stages, including lab tests on grain biochemistry, and field trials of hulless oats. Once promising varieties are identified, they need to be grown out to scale for the feeding trials with live hens that will be carried out by Oregon State University animal scientists. The oat-based ration will be evaluated for its nutritional value to laying hens, and the hens' eggs will be evaluated for their nutritional value to humans, to see if feeding oats to chickens changes the health properties of their eggs.

If this research demonstrates that hulless oats could be valuable for poultry farmers, sourcing the right grain might be difficult until more grain farmers begin to grow it. Not many farmers are growing hulless oats in western Washington, and even if they wanted to, seed of specific varieties is not necessarily easily available. The pacing in plant breeding is not exactly glacial, but it is definitely slow.

No one's told the chickens that better rations are on the horizon, though, so they are as patient as the researchers—and anyone in the middle of agriculture who understands how long it can take for an idea to be realized.

The hunt for new oats is looking for other characteristics, too, such as suitability for fall sowing. Nearly 80 percent of the oat crop in the United Kingdom is planted in the fall, but where oats grow in the United States, spring oats are the rule. Fall oats can offer higher yields, better weed control, and winter ground cover, the same as with any other small grain.

Louisa would like to see an entrepreneur fall for oats, and start a small oat-handling operation in western Washington offering a locally grown, locally milled product, maybe even using named varieties. Given the strength of the gluten-free market, the prospect seems plausible.

From the mid-nineteenth century through the mid-twentieth, oats were a high-yielding cash crop all the way down the Salish Coast. They may never reach those heights of popularity again, but the potential for reintroduction is real, and just needs some investigation. Opening up markets for oats as chicken feed or specialty food products might help western Washington farmers to get more value from grains, rather than seeing them as a dead spot in the rotation.

When I first met Steve Jones, I thought he was a figurehead. He was pouring the Kool-Aid for all kinds of people, from grad students to farmers and grains fans like me. His drink was mine, of course, so I was smitten with his life and work.

I am no less smitten, but I see that this is not a cult of personality. He is making great things happen in grain production in regions both near him and near me, having landed in the right place at the right time.

The work in regional grains at Mount Vernon is taking roots and fanning out, training people who are making other projects possible, like Julie Dawson, who has been involved in grains projects in the Northeast and in France. Steve consults with people in Vermont, New York, North Carolina, and California, lending his input on regional variety selection and other matters to mills, restaurants, and bakeries.

The model that is developing in the Skagit Valley is an argument for localized food production that goes beyond the lore of local eating. This is not about romancing a precious philosophy. The goal is to keep farmland productive. Development pressures in the I-5 corridor are extreme. If people can raise the economic profile of commodity crops in such a competitive environment, there's great hope for changing approaches to grain anywhere.

At its new location, The Bread Lab has kept its flour-centered name. The work has never been limited to bread; the place is a resource for studying all grains, as well as many grain-based food and beverages. Washington State University is raising money for the fermentation side of the equation. A staff fermentation expert will come on board, and a full brewing laboratory is planned as well.

Soon people will be able to enroll in a certificate program at Skagit Valley College and learn how to brew at The Bread Lab. The port is thinking of creating a brewing incubator, a place where brewery start-ups could launch and grow their businesses.

The Bread Lab is part of a national conversation on grains, and companies like Chipotle Grill and King Arthur Flour are involved in the dialogue with both words and money. The flour company formalized its interests in The Bread Lab with a partnership that includes King Arthur bakers regularly traveling to Washington. The purpose is educational, and the information flows in many directions, between researchers and bakers, and from the bakers to professional and home bakers in classes.

Jeffrey Hamelman, head of the King Arthur bakery in Vermont and a widely respected authority on bread, is a presence. His classes in Mount Vernon fill up in minutes. In his keynote speech at The Grain Gathering, he talked about the importance of learning, for students to give themselves permission to fumble, and for teachers to be conduits for information and experience, committed to opening doors for others. He extended that consideration to the current moment in grains, when learning about bread is occurring across professions.

"As bakers, it's no longer enough to just think about a loaf," he said. Bakers are expanding their thoughts and work, connecting with farmers and millers. "We're each holding our place in the traditions. When we eat bread we absorb the labor of so many people."

Farmers, he closed, are the biggest risk takers of all.

CHAPTER TEN

SEEDING NEW GRASS

*W*hat I love most about stories is the seeds they plant for change. The 100-Mile Diet, the concept that coined the word *locavore*, began when a couple in British Columbia decided to eat from a tight radius around them. The story rippled out first in articles and then in a book, helping steer people toward the same practice.

When word of the diet hit a co-op newsletter in Oregon early in 2006, the wheels of change began to turn. Co-op member Chris Peterson checked the bulk department and found that none of the staples were locally grown. She wrote her own story for the newsletter, daring fellow members to think about changing the contents of those bins.

That was a challenge that Harry MacCormack could embrace. The pioneering farmer, co-founder of Oregon Tilth and author of early standards for organic certification, bought sixteen small bags of organic but anonymous commodity staples. Black beans from China were the best performers in the test plots at his Sunbow Farm; that these beans, products of industrial agriculture, did best for this local farming project was ironic.

The Willamette Valley is home to some fairly enlightened eaters. These people live in the cities of Portland, Eugene, and the capital, Salem. Oregon State University, the state's land grant institution, is in Corvallis, and serves natural resource industries like timber and farming. Despite the valley's relatively dense population, commercial agriculture is an economic driver in the area, which is known as the Grass Seed Capital of the World. In addition to grass, plenty of other things grow here, such as hazelnuts, Christmas trees, and wine grapes. A lot of land is planted with clover, meadowfoam, and mint. Edible grass plants like wheat are grown in rotation with many crops and sold as commodities, mostly to Asian noodle markets.

This is a diverse farming region, but over the last quarter of the twentieth century production of seed crops for grass, flowers, and vegetables began to dominate food crops. In the 1990s, a number of interests converged to study food access in a strong farming region. Dan Sundseth, who worked with the Farm Service Agency, became the USDA food security liaison for the state through the Community Food Security Initiative. This initiative was not funded, but identifying the issue and giving a title to someone like Dan, who worked directly with farmers, was enough to get the ball rolling. He connected farmers with community activists in the Willamette Valley and started gleaning projects. People began looking at other ways to address the flow of food in the area, and such considerations included the topic of farm biodiversity. In 2006, questions about farming and food access narrowed to staple crops, with the Ten Rivers Food Web and another group, the Willamette Farm and Food Coalition, leading the dialogue. The group applied for and got a grant from the USDA to test growing beans and grains on a larger scale than Harry MacCormack's initial plantings, and the Southern Willamette Valley Bean and Grain Project was born. Because of Dan Sundseth, the project gathered farmers from either side of the conventional and organic divide.

Willow Coberly, founding member of the Bean and Grain Project, lives right on that divide. Having grown up eating brown bread sandwiches and backyard vegetables, organic is a natural part of her thinking. But her husband, Harry Stalford, is from a very different mind-set. The family business, Stalford Seed Farms, is a model of successful conventional farming.

Willow and her mother, Gian Mercurio, began to try to grow organic food at a farm scale in 2001. However, any weed that came up got hit with a spray until 2004, when they brought home facts from the first World Organic Seed Conference in Rome. Armed with information, they convinced Harry Stalford to convert a portion of their acreage to organic. Willow and her mother got the two Harrys, MacCormack and Stalford, talking about soil fertility and other farming issues. These conversations helped farm and develop 11 acres they had in beans and other vegetables.

By the time of the USDA grant, Willow and Harry's farm was the right place to trial more staple crops. Growing at Harry MacCormack's Sunbow Farm would be much less visible. As esteemed as he is in the organic community, what happens on his farm means little to conventional growers. Harry Stalford, though, is a respected grass seed farmer, a field that has zero to do with organics.

Farming is a very public way to earn a living. The fact that Harry Stalford was willing to take a risk and grow something chemical-free was huge. The risk was bigger than the money those acres might or might not yield. Harry and Willow gambled with their standing in the community. If they could organically grow, sell, store, and distribute grains and beans in the region, the concept might have legs.

Inside and outside of the Bean and Grain Project, people watched. The group gained momentum and members. Twenty and then thirty people gathered monthly to discuss techniques for growing and marketing staples, continuing to draw organic and conventional farmers to the same table. The meetings also broke down barriers within those sectors, encouraging communication between would-be competitors who shared markets but normally guarded their interests.

This was practical and radical, a community-driven motion to change commercial practices. The proposed revision began with idealistic intentions, to serve the health of the soil, the people, and local economies in the face of climate change and peak oil. Then the recession hit, and the radical idea began to seem practical.

Housing starts stopped, and grass seed quit flowing out to the world. The time seemed ripe to convert from an economy based on turfgrass to other farming options.

Change may strike like lightning, but adaptation is not as dramatic. Here's how the changes looked at Harry Stalford and Willow Coberly's farm. Their investigation of organic food production seemed prescient. The totes of organic grains and beans they'd grown took on a new glow in the warehouse, sitting like cash beside sacks of suddenly unwanted grass seed.

Willow started Greenwillow Grains in 2009. About 100,000 pounds of grain each year are processed in a space no larger than a shed. Workers mill and sift bread and pastry flour, and buckwheat, rye, and triticale. They make rolled oats, too. Customers are local and loyal, and all the grains come from

Willow and Harry's farm, which now has 500 certified organic acres. In any given year, about a third of that is planted to wheat.

"I'm getting calls from all over the country for quantities I can't even fathom, because we can pinpoint the field and guarantee organic and non-GMO. I hate saying no," said Willow. She would like to convert more acres, but the transition years are economically tough. (Organic certification requires proof that land has not been chemically treated for three years. Crops planted during that time can't be sold at organic premiums.)

While she wants the mill to grow, she doesn't want to run a huge facility, either. A USDA Value Added Processor grant will enable an expansion of the mill, in both infrastructure and hiring. Putting people and incomes into food processing is important to Willow. The idea of grains from the valley heading to Asia on ships, and passing ships that are delivering food to Oregon tables, vexes her. Yet this reality remains intact, partly because organic beans, for instance, are cheaper to buy from international suppliers. That price differential limits the market for regional foods.

As long as cheap conventional commodities and less expensive organic staples exist, most of us don't want to pay the true cost of food. We've grown used to food that doesn't reflect the costs of production, some of which are hidden behind subsidies and futures trading. The reception of regional grains is further narrowed by a battle between organic and local. In some cases, shoppers value regional farming over organic labels. For the mainstream consumer with a preference, organic trumps local, simply because it is more available and often cheaper.

The Willamette Valley hasn't transformed overnight. In 2008, the state's grass seed income topped $500 million. While the figure has since dropped to $300 million, staple crops are not suddenly dominating production. At first glance, this might seem puzzling. But the disappearance of one market doesn't make another magically appear. Having the tools to handle inedible grass seed doesn't mean you can just swap seeds and start handling edible relatives. The quality parameters for food production make this more than a lateral move. Aside from that, the basic issue of soil is a determining factor.

Typically, there are many types of grass seeds grown in the valley: annual rye, tall fescue, perennial rye, orchard grass, bent grasses, Johnson grasses, and many others. Annual ryegrass will grow in waterlogged clay soils, but the other grasses won't, and neither will wheat. You can encourage growers with clay soil to plant wheat, but they won't do it, because they know they'll lose their crop. Aside from the suitability of soils, there is price to consider. If you have ground good enough to grow a high-value crop like berries, why on earth would you plant wheat?

Since the recession, some grass seed farms have disappeared, and other markets for turf seed have opened up. China became a major buyer. Most of the wheat grown in the valley is still the soft white type, grown for the export market.

Even if the conversion was not enough to satisfy dreamers like me, stunning changes in grains have happened in Oregon. Another, much larger mill opened, Camas Country Mill. And Hummingbird Wholesale has pioneered a trend that's worthy of notice and replication: distributor-supported agriculture. The evolution of the mill overlaps with the distributor's interest in developing regional farming.

Tom Hunton, farmer and founder of Camas Country Mill, and Julie Tilt, owner of Hummingbird Wholesale, were both at meetings of the Bean and Grain Project. Theirs is an unlikely partnership, straddling that conventional/organic divide.

Julie and her husband, Charlie, bought Hummingbird Wholesale in 2003. The company began as a honey shop in the 1970s, then shifted to natural foods distribution in 1981. Hummingbird Wholesale distributes to natural foods stores and restaurants between San Francisco and Bellingham, Washington, close to the Canadian border. The couple believes in change the way that most people trust the clock. In the dozen years they've owned the company, they've made great strides in building an environmentally responsible business. Investing in a regional food system is part of their design, and that's why they were at the meetings about staple crops.

Tom Hunton is another respected conventional farmer in the Willamette Valley. He and his son farm more than 3,300 acres, just as he did with his father. The family farms a range of seed crops, including grass, clover, brassicas, and some grains. Prior to opening the mill, many of the grains went for export. Tom still runs the fertilizer distributorship he started with his father, Sure Crop Farm Services; this shows that diversifying enterprises has long been a habit in the family. So when the grass seed market hit the skids, Tom was well positioned to find alternative income streams, and Tom approached Charlie and Julie about growing food for Hummingbird.

Tom began planting bread wheat, trying several kinds to see what might work. The wisdom in the area was that hard wheats with acceptable protein levels for bread baking could not be grown, but Tom had enough faith and farming experience to try. The crop was successful, but the cost of having it milled nearly 200 miles away, at Butte Creek Mill, stole any potential profit. Charlie and Tom began having conversations about building a mill. Another person in on those meetings was James Henderson, Hummingbird's new farm liaison.

James's job is to be a bridge between farms and markets. All distributors are in this position, but the two worlds of farming and marketing don't always share a language. Charlie and Julie wanted to develop local foods that they could sell, but they were pinched for time. They also were at a disadvantage because they were not from the farming community. Hiring James, who was a farmer, was like hiring a translator who could articulate their goals to farmers. More important, as a ninth-generation farmer he had an insider's understanding of agriculture.

"We had lots of good ideas but no time to implement them," Charlie explained. They had their hands full with any number of projects and more than thirty employees. The business they bought had 1,200 square feet of space, and they were in the process of expanding their headquarters to 19,000 square feet. They chose to be near the city of Eugene so they could best facilitate change in the community. The building hosts an incubator kitchen for local food enterprises; other space is leased to businesses with similar values-based goals. Within Eugene, deliveries happen via cargo bike. *Zero waste* was the policy followed as they outfitted the building, and it remains a primary goal.

James Henderson first interviewed for a job as warehouse manager, but Charlie and Julie created a new position for him once they realized his qualifications could help facilitate the strengthening of the local food system.

As farm liaison, one of his first tasks was researching the mill. He interviewed bakers to see what they'd purchase and how they needed to have flour delivered, how often, what size bags, and how many of them? James, Tom, and Charlie met weekly as a team organizing what kind of mill the area could support, and scouting locations. The process took a year and a half, far longer than anyone had anticipated. Identifying what kind of mill was simple. Bob's Red Mill, a long-established whole foods operation near

Portland, used a certain kind of Danish stone mill, and so many of them that Tom didn't consider another brand. Building permit hurdles were nowhere near as easy, delaying the project and adding an extra expense—$30,000 for a dust control system—to an already significant investment of $300,000.

Tom and Sue Hunton financed the mill from their savings, an economic development grant from Lane County, and a small loan from Hummingbird. That loan is now repaid, and the mill has become a fixture in the community.

Like other civic structures, Camas Country Mill is a tool and a symbol. The first in the valley for eighty years, the mill grinds regionally grown grains, providing critical infrastructure for farmers. Tom and his son Jason grow crops for the mill, a mixture of grains ranging from barley to teff. Some of the crops are grown organically, and some are grown on land that is in transition. Some of the flours are custom-milled for Hummingbird Wholesale.

Camas Country Mill has another role, acting as intermediary between farms and eaters. In this capacity, the mill provides whole-grain flour for seven school districts and a lentil-barley soup mix distributed by the food bank. These are connections that go beyond farm viability and show why farms should feed the communities around them: not just because it feels good, but because the health and strength of communities weaves together seemingly disparate populations, economies, and needs.

The work and impact of the mill are not limited to Oregon. Camas mills teff for the Ethiopian and Eritrean diaspora in the Northwest, and fields requests for teff from around the country. Camas also mills barley flour, which is paired with teff in the making of *injera*, the flat bread served with traditional Ethiopian dishes. These buyers have an interest in chickpeas, too, which pleases Tom because it makes a good market for necessary crop rotations.

Camas Country Mill shows how a diverse line of grains and a diverse network of growers can collaborate to generate forward motion on the dual tasks of farm and food security. The mill's beginnings with Hummingbird demonstrate how partnerships can drive change and develop markets. The wholesaler buys many of the mill's products.

As other regional grain systems grow, the mill is a model, and also a funnel, able to offer Oregon grains into burgeoning systems. In Los Angeles, Grist & Toll is purchasing some heritage grains for milling that are not available in Southern California. Colorado restaurant Pizzeria Locale is using grains from Camas as it mills flour on-site for long-fermented pizza doughs.

One of the most gratifying functions of the mill for Tom and Sue Hunton is its educational capacity. Since the mill opened in 2011, they and one of their employees, a former educator, have been teaching kids about grains. Kids grind grains with stones and bricks to get a firsthand experience of milling. They see the machinery and follow the flour from fields to food, sampling baked goods and bringing home mixes to make for their families. This educational component is such an important part of the mill that the Huntons have purchased and are restoring a one-room schoolhouse. This will be a space dedicated to the more abstract civic notions their business performs. Through such educational efforts, that string of connections among farms, markets, and consumers is made visible. Farming's central role is highlighted.

Hummingbird Wholesale, too, highlights that role, negotiating its position as a distributor to create a robust local food system. By having a philosophical stance and backing that stance with staffing and other mechanisms, such as buying seeds or helping build necessary infrastructure like a mill, the company is transforming options for farmers and options for consumers.

"Markets don't demand what they don't know exists, and farmers aren't going to grow without the market," said Charlie. Consumers can vote with their forks all they want, but those votes need to be back-loaded. "We're uniquely situated in the supply chain because we can make the connections between farm and consumer."

Pumpkin seeds are his favorite way to explain these ideas in action. The company worked for years with farmers in the area to grow hulless pumpkin seeds, and the crop is a great success for growers, and for the company. The variety is native to Styria, Austria. Styrian pumpkin seeds grow well organically in the area, and now the state is the leading producer of these pumpkin seeds in the nation. One grower had such good luck with the crop that now he's marketing it on his own. This is a point of pride for Charlie, not a loss. The pumpkin seeds are still a great seller for Hummingbird, and demonstrate the kind of opportunities the company is trying to share with farms. That the crop does well organically is another plus.

"We're changing the agriculture," Charlie said. "By offering a different price and new crops, there's a cascading effect that occurs. We want a working system that has four to five crops that can be rotated. We're trying to build programs for transitional lands so they have a one-stop marketing opportunity, and not just have something to sell one year."

The rotations are going to be different in eastern Oregon than in the Willamette Valley, but either place requires multiple farmers so that each year the rotations are marketable. One farmer grows chickpeas while another is growing black beans and another is growing wheat. This is a complex and uncommon thing to engineer, but Hummingbird is used to working outside the bounds of tradition. Just keeping the warehouse stocked with a year's worth of regionally produced goods, rather than buying as needed from commodity supplies, ties up a small fortune.

James Henderson's job is a key part of this pioneering route to better farming. Nine generations deep in American farming, he traces his Scottish ancestry through Virginia. He farmed in the 1990s and worked for Weyerhaeuser at night, and jokes that he shut down every lumber mill in Oregon during the downturn in 2008. His territory for growers or buyers is not limited to the Willamette Valley, but his work focuses on staples, not produce. He builds relationships with growers and customers, talking and walking crops into the field and out to the user. He buys seeds, writes contracts, and finds other ways to facilitate production.

Take the case of wild rice. Because of Hummingbird's purchasing power, one wild rice farmer took a big leap and bought a parcher and dehuller. This aquatic crop can grow on otherwise unproductive ground, but prepping that ground with berms and dikes takes labor and dollars. Processing the crop is demanding, too. Wild rice is harvested semi-green; dry kernels shatter when they hit the combine. The parcher cooks off excess moisture, and the dehuller removes the hull. These are tools with big price tags, and contracts from a steady customer make financing such investments seem reasonable to a lender, not ridiculous.

Another way Hummingbird supports regional grains is finding secondary markets, as James did when contracting for food-grade soft white wheat.

"If the wheat doesn't meet the standard, what am I going to do, just tell the farmer, oh well, tough?" said James. He felt obliged to find another market. The pricing is strong enough for feed-grade wheat that some farmers contract strictly for that, and don't have to worry whether or not the crop will make the food grade.

The company gives James room to expand the market like this, and develop opportunities, like the Styrian pumpkin seeds. A couple of barleys on the sales list illustrate the collaboration that's possible when money meets ideas.

Streaker and Karma Purple are hulless types of barley developed at OSU's Barley Project. Barley breeder Pat Hayes explores varieties in test plots, but the foods only reach consumers through commercial partners. Tom Hunton was willing to take a chance on these barleys at his farm, in part because of Hummingbird's ability to sell new foods. Both barleys are high in beta-glucan, a water-soluble fiber, and have a low glycemic index. Their deep colors comes from anthocyanins, which are antioxidants.

In his sales capacity, James told the stories of these barleys, including Karma's Tibetan lineage. His excitement didn't immediately catch on, though, and the Karma barley sold slowly at first, only about 25 pounds a week.

"Then all of a sudden, zoot, I don't know why, but we're taking off and we're selling 1,000 pounds a month and it's continuing to increase," he said. Who knows what the tipping point was, but somehow the nebulous fork of the consumer finally speared this product. Hummingbird could only load those forks because of the legwork that it and others had done for the barley. We fill our plates thanks to markets and infrastructure, not just wishing for better food options.

James Henderson does a job that many farmers do not want or have time for. He sells crops. This is not just a convenience. I've watched farmers juggle sales, and bakers, millers, or maltsters shop for grains. Establishing relationships is a time-consuming second or third job. Getting to know James and his work, I see a place for grain brokers who specialize in buying and selling unusual grains. The idea of the middleman doesn't jibe with popular notions of independent small-scale farms, but direct sales don't work for everyone. Farmland is at a remove from population centers, and grains require volume production to balance the costs of infrastructure, maintenance, and land. Middlemen are essential to revising the way we grow and get staple crops. They can keep people connected, and free up their time to focus on farming or making food.

James reminds me of Jesse Hawley, the flour merchant who imagined and argued for the Erie Canal. Hawley worked in the early 1800s, when knowing your farmer was not an advertising ploy. If you were a baker or grocer in New York City, you had to trust your liaison to get sound wheat upstate, and have that milled and shipped in a reasonable fashion. Few farmers, then or now, could sell just to their neighbors and thrive. Hawley and other middlemen connected people and products. Aside from the financial incentives that Hummingbird offers to support new options for farmers, James's work as a specialty broker seems critical in helping people work off-grid in staple crops.

"Without the market, why would you plant it? The price has to make sense," said James when I asked him to help me understand how markets leverage change. Farmers have to make more than they could growing for feed or selling for exports. "Once you do that, a lot of things can take care of themselves."

James is watching things take care of themselves bit by bit in the small towns where he works, places that might have five hundred or seven hundred people and have been left behind by timber or other industries.

"I can come in there and contract for wheat, or barley, or garbanzo beans, and get that market flowing," he said. The economic impacts are evident in little things, like someone driving a new truck, or new siding on a barn. This is how rural America is going to survive, he said.

The sales side of his job creates new opportunities for connection, too. Through James, bakers can follow a bag of flour all the way back to the field. One way James tells the story is through pictures he sends, of farmers and their wheat. From tags on a bag, bakers might make a link between the flour and Tom Hunton's wheat, for example, or Tom and Ray Williams's wheat.

"They get this mental connection with people they've never met," said James.

James can also illustrate the grain's trail through money, tracing the costs. The farmer gets this much, and the miller gets this much, and the bag costs X. Transparency is important as people weigh the price of regionally made goods, so that they understand where the money goes in a supply chain based on values, not just price.

Enumerating those prices reveals the footprint of regional spending. When New Seasons Market, a small chain of groceries with bakeries, started buying a good volume of flour from Camas Country Mill through Hummingbird, nearly $300,000 stayed in the region, and more than half that went to farmers. Milling, seed cleaning, and bag manufacturing businesses got paid, too. The amount Hummingbird collected for arranging the logistics and the cost of delivery is small, especially when you consider how thin margins are for food wholesalers. Paper-thin.

Those Styrian pumpkin seeds are another good demonstration of the two sides of James's work. He is able frame the crop as a good opportunity for growers, and the food as good for buyers.

One of Hummingbird's clients is Tabor Bread, a bakery in Portland. He sold manager Annie Moss on the pumpkin seeds with his enthusiasm about the taste. His ability to describe the plants in the field and their beautiful giant heads convinced her to try them.

"Pumpkin seeds—who gets excited about pumpkin seeds?" she asked me. What she thought of as a pedestrian food has surprising flavor. The odd thing of this is that the bakery where she works is all about bringing surprises to standard stuff.

Tabor Bread lives in a house with a peaked roof at the end of the Hawthorne District, and at the base of Mount Tabor, the cone of a dormant volcano. The location straddles the Pacific Northwest combo of quietly urban and bold natural settings, which seems fitting, since bread is a bridge between nature and food, between people and land.

You can sit in the small bakery and feel at once a part of this place, yet at the same time ready to swing back into the groovy and subdued hubbub, or up to the park. When I visited I did both, sharing lunch and then a loaf of bread with a friend. After my friend left, I took a walk, circling the reservoir and threading through drapey evergreens up to a sweeping view of city and highways and trees.

The bakery opened in the fall of 2012. The kitchen is well hidden, but the mill room has windows, and the oven is a part of the café, a small monster that's far more eye catching than a chandelier. Behind a counter stacked with sweets and savories, bread sits on racks. Clerks can answer intimate questions about the foods and drinks, and there is another counter with stools and a full view of the oven. You can watch the baking like you're at a sushi bar.

Their biggest bread is the Fife Boule. I was naturally attracted to this bread, which is neat because this is the loaf that launched this bakery.

"I wanted to create something, and it started out with a loaf of bread," said Tissa Stein, a reserved woman with long white hair. The loaf was a Desem loaf, the ideal Alan Scott loaf of bread. She met Alan and the bread more than twenty years ago, and after raising her family and having a few careers, she was ready to make something. The bread led her to a solution.

Tissa found Alan Scott, the oven builder whose designs launched micro-bakeries across America, when she was living in Petaluma, California, and on the lookout for good food. Once a week, she drove to get four loaves of his Desem, bread that was of such a different category, her kids called it by the baker's name.

"I never had anything like it," she said. "There was bread and there was Alan Scott bread, and we all knew what we loved."

Eventually, she asked him to build her an oven on her ranch in Petaluma, California. Tissa made pizza, and played with Alan and Laurel Robertson's recipe for Desem. The oven was open to friends as well. When her friend Jed Wallach, an artist, became quite taken with bread, he started baking once a week and selling what he made at the end of the driveway. Soon he opened a bakery in Freestone with a large Alan Scott oven. The place, Wild Flour Bread Bakery, is a destination, baking nine hundred loaves of bread four days a week.

Tissa's own route to bread was indirect. When her husband passed away in 1992, she turned the ranch into a horse boarding business, and took other jobs, too, including working with teenagers in the judicial system as a coach. When her children were grown, she took the chance to travel and live in Europe. That's where the itch to make something occurred, and the idea of opening a bakery sprouted. She considered starting a project there, but she didn't want to be so far from her family.

So she found Portland, not because she was scouting a hot spot for bread, but circuitously, following her interest in tango dancing. Discovering there was no wood-fired bread bakery in Portland, and no bread made from house-milled flour, she realized that she'd found the ideal spot to launch her idea, smack in the middle of a community with a good interest in food.

As she planned her bakery, Alan Scott's voice was very much with her. His daughter Lila is carrying on the legacy of her father's ovens and ideas, and Tissa consulted her as she designed Tabor Bread. Alan Scott had a strong concept of what it meant to be a village baker. He insisted on fresh-milled flour, and direct-fired ovens of a certain size; limiting the size of the oven and fire would limit the amount of bread produced. Once a community needed more bread than the baker could manage, a new bakery should arise.

Tissa chose an Osttiroler mill, which was Alan's recommendation, and picked out the type of mixer he liked, too. For the oven, she chose a next-generation model designed by bakers who had bumped into the ergonomic and economic constraints of the original Alan Scott design. In practice, village bakers have found they needed to make more bread.

Andrew Heyn, of Elmore Mountain Bread in Vermont, worked with mason William Davenport from Turtlerock Masonry Heat to build a larger brick oven. Davenport also built Tissa's oven, and she opted to include an external firebox, too. This is a slight deviation from Alan's dictate to heat the baking chamber with a fire, but Tissa wanted the option of adding additional heat to a bake if necessary—for instance, if they wanted a heat bump to make baguettes. The firebox is not frequently used, but the flexibility is there. This is an example of how compromise has to live in real-world expressions of principled pursuits.

The same compromise came into play with that Desem loaf, and how they described it. Made from half heirloom Red Fife wheat and half modern hard red wheat, the bread is earthy red, but also gray and brown. This feels like food you could really live on, the stuff that gave the staff of life its name, and harks back to times when people got most of their calories from bread.

"When we first opened the bakery we were really excited about that bread," said Annie Moss, the manager. "People would be excited the grain had a name."

However, the title Fife Boule took too long to explain, and by the time people got the grain's story, their attention had evaporated. People just wanted a loaf of bread. So the story got condensed to fit in a quick exchange, a simple description of flavors and colors.

The baseline knowledge of grains in the general public, even among people who are led to real bread, is limited, and the staff at Tabor Bread is always learning how much information they can feed. While awareness is growing, people don't know the basics about whole grains. We live in a country that doubts the edibility of a foundational food, so to find anyone's point of entry, you have to be careful and responsive. Maybe today you can talk about gluten breaking down during long fermentations, and in six months, you might be able to have a conversation with the same customer about different varieties of wheat, and their various types of gluten.

Tabor Bread is creating opportunities to discuss the differences of the grains and bread at length. They had a series of suppers, one for Red Fife, one for rye, and one for buckwheat. Kathryn Yeomans, a chef who cooks at farmers markets, made four-course meals featuring the grains. Growers came in two capacities, to enjoy the dinners as guests, and to narrate their experiences. A rhubarb farmer came to the buckwheat dinner, too, because buckwheat and rhubarb are related.

People came from Camas Country Mill to share the meals and share what they know and love about these grains. Tom Hunton told his version of the legend of Red Fife. Elizabeth Hayes brought her insight on rye. Annie said she could see people making connections between what they ate and where it came from, awareness flashing across their faces. She was thrilled to see the exchange.

The bakery doesn't make a million kinds of bread, just a handful of standards and a daily special. This limit simplifies production, and also allows the grains to introduce themselves. The batard is the most quiet bread. Made from white whole wheat flour and white bread flour, the grains don't have a lot to announce. The other breads are platforms for the spelt, rye, Red Fife, and other grains to announce themselves.

Camas Country Mill provides most of the whole grains the bakery mills and sprouts, but Tabor doesn't mill all its flour. Pastry flour comes from Camas, and white bread flour comes from Central Milling in Utah. Fairhaven Mill in Washington grinds the buckwheat. The bread and its stories reach people at farmers markets, another great stage for learning about grains. Many people meet the loaves at less chatty but higher-circulation places, like food co-ops and Whole Foods Market.

Annie has a nifty grains story herself. She was June Russell's assistant at the Greenmarket Regional Grains Project, and fell into the bakery when she came home from New York City and graduate school. She was looking for a food-oriented job, but thought she'd be working more generally, advocating somehow for regional food systems. When Annie saw that a bakery was going to be using all local grains, she broke free from her usual cover letter to say she didn't have a restaurant background, but "you should know I exist."

Annie fit perfectly, because the team was forming around an idea, not around careers in food. Tissa, at sixty-four, was building the bakery with Corey Mast, who at twenty was already a solid baker. Their goal was to make a spot where the traditions of fresh milling, wood-fired baking, long fermentations, and wild yeast could coexist.

"Our reference has always been finding our way back to as much whole grain in a loaf as we can," said Tissa. She finds the rhythms of her other work lives very present. Baking bread is remarkably similar to boarding horses. "I knew how to be responsible for a barn full of animals. Being responsible for a bakery, and who is going to feed the starter each day, and tend the loaves, is familiar."

Tissa is more of a recluse than the job allows. While I've seen bread as a route to community, and bakeries as a hub where people connect with one another and with food, Tissa's intention is more about building a community within the staff.

"You're cultivating sourdoughs and sort of cultivating this other culture inside the bakery," she said. Whatever unity the bakery has will spill out to the customers, and she's worked hard to make the environment support-ive—and somewhat soft, even against the time pressures of baking and the inevitable personality crunches that are part of retail shops. She is pleased with the relationships and resilience that have developed among the fifteen people who work with and for her.

The original baker, Corey, has moved on, but his stamp on Tabor Bread lingers in the breads and in the relationships he built with local growers. After more than two years at Tabor Bread, Annie moved on, too. She felt the moment was right to start her own grain venture, with Katia Bezerra-Clark, a longtime friend and coworker from Tabor Bread. The two teamed up with Handsome Pizza, an existing wood-fired pizzeria, to open Seastar Bakery and Handsome Pizza in the spring of 2015. The enterprise has its own mill.

At Tabor, the signature Fife Boule holds their stories and more—more stories than it can tell. Made from Oregon grains, it holds the achievement of the Bean and Grain Project, including the 1,000 acres in the valley that grow organic staples, and the two mills. The bread can't articulate how those changes happened, but people who are interested in changing food systems look to archived records online. People also contact Willow Coberly and Harry MacCormack for advice.

The Fife Boule speaks of other practical, radical thoughts, like Alan Scott's ideal of the village baker. The taste is haunting and solid, and chases me like a ghost and a friend.

CHAPTER ELEVEN

VALLEY MALT

𝓝ot many home brewers end up growing 40 acres of barley, let alone malting 5 tons of grain a week. But a casual hobby took on epic proportions for Andrea and Christian Stanley, who built Valley Malt, New England's first malthouse in a hundred years.

The town of Hadley, Massachusetts, is flat, and the roads cut like lines through a grid of fields and houses. Crops run flush with backyards. Forward-thinking food activists are longtime residents in the Pioneer Valley, which is home to Amherst, Smith, Mount Holyoke, and Hampshire Colleges and the University of Massachusetts; graduates often settle here, along with their ideals. Food businesses like the South River Miso Company began in the 1970s, and sustainability was a community practice in this area long before the concept of local eating hit the tip of the national tongue. Hadley seems the perfect place for malt's reentry in the Northeast.

Just as flour needs to be milled for bread, grains need to be malted for alcohol. Malting cracks open a kernel's gold, priming the starches so they can convert to sugar and feed fermentation. Back in the days when beer was a local product, many breweries had their own malthouses. But agricultural drift swept grain production out of sight, and Prohibition and other regulations squelched local brewing and malting.

Late in 2009, Andrea, a social worker, and Christian, a mechanical engineer, read that farmers were growing grains nearby. They immediately thought of putting those grains to use in brewing, maybe even in their own brewery. The step of malting was so invisible they didn't consider that getting grains malted would be a hurdle, let alone a fence they would have to climb.

Quickly, they discovered that most malt comes from large facilities. A few small-scale malting operations have been around in England for five hundred years, and in Germany for even longer, but malt in America is Big Malt, where the smallest factories make 150,000-pound batches. The last malt in the Northeast was probably made in Buffalo, on the western end of New York, where Christian is from, or at Rochester's Genesee Brewery. Grain processing lingered there because of its location on the Great Lakes. Now most malt used in American beer comes from the Midwest, western states, or Canada. Equipment for malting on anything but an industrial scale does not exist.

While I could spend a lot of money on a fancy countertop flour mill, there is no equivalent for malting. Christian and Andrea researched the lost craft, and he designed a prototype malt system.

"When I was in research and development, I worked with PhD scientists on projects," he said. Christian is thin and blond and in his late thirties. He speaks quickly, moving words as efficiently as he works. Father, farmer, maltster, and mechanical engineer, he has a lot of tasks to tackle.

"The scientists would say these are the conditions we need to create for these processes," he said, explaining his facility with materials. "I knew that I could build a system that could malt grain, but I didn't know what the parameters were for designing that system."

He built a prototype to figure out some of those parameters, installing a control system to regulate temperature and airflow.

Malt took a fierce hold on their lives, really staking a claim on Andrea's imagination. She befriended librarians and historians, immersing herself in the global and local history. Until the Industrial Revolution, malting was a domestic chore. The first white settlers to Hadley malted in their kitchens, same as they made bread. She found the names of barley types that once grew nearby, and the names of the early professional maltsters. She and Christian visited Malterie Frontenac in Quebec, one of the few micro-malthouses in North America. They wrote a business plan. They talked to Jack Lazor and Heather Darby in Vermont about grain growing in general, and what types of barley could grow in the Northeast. They called the American Malting Barley Association for advice on varieties, too. The question had not been asked before.

The tabletop prototype could steep, germinate, dry, and kiln 10-pound batches of grain, so they fiddled with barley like mad chemists, malting and testing the results by brewing beer. They scouted a location, farmers, and seed. The Northern Grain Growers Association invited them to do a presentation at its March conference. Andrea and Christian hauled their prototype up to Vermont and shared what they knew, wondering if their pursuit was totally nuts.

The wheel of malt kept spinning around their busy lives. They had a new baby, their third child. Christian worked full-time as a consulting engineer, and Andrea worked part-time. She signed up for a short course on malting barley at North Dakota State University. Farmers planted spring barley for them. They found a suitable space, a two-bay garage that used to be a small potato-processing barn. They rented half of it and came up with a name, Valley Malt.

Christian scaled up and tweaked his designs for the 1-ton malting system. They had the stainless steel cut and rolled at one place, welded at another, and ended up with a vessel that could handle all steps, from steeping to kilning. Christian rigged electronic controls to monitor temperatures and moisture levels, keeping track of the grain's environment at each stage. They hired a plumber and electrician, and installed the malt system. Other equipment took over the garage at their house: a seed cleaner to remove field debris and weeds; grade kernel sizes; and a roaster, to add extra flavors.

Here is how the grains transform.

The first step is soaking or steeping the barley or other grains in water for one to two days in two or three repetitions of water on or off the grain. The goal is for the kernels to reach 45 percent moisture, on a time line determined by kernel size and the initial moisture level of the grain. Once the desired hydration is achieved, the germ in each kernel is ready to sprout. The tank is drained, and a blower is hooked up to push air into the grains at about 60 degrees Fahrenheit (about 15.5 degrees C). Within twenty-four hours of the initial soak, germination begins, and it continues for another few days. All the while, the blower circulates air through the grain bed. The maltsters stir the grain every twenty-four hours, breaking up rootlets and preventing the formation of brick-like chunks of malt. Some systems have automatic stirrers, but most micro-maltsters get in the bin and move the grain around by hand.

Kilning or drying is the next step. This means dehydrating the grains to stop germination. The temperature has to be between 110 and 115 degrees

Fahrenheit (43 to 46 degrees Celsius). The idea is to halt the physiologic processes at just the right moment, when enzyme activity in the grain is poised to shift the starches in the endosperm into sugars. *Modification* is the technical term for this process, and malt analysis tests at the end of malting show brewers how much modification has occurred.

The first stage of kilning blows a lot of low-temperature air over the grain bed, because higher temperatures will denature the highly desired enzymes that the grain's brain, the germ, has triggered to release. After most of the moisture is removed, the enzymes are not as delicate, and the temperature of the tank can be raised more quickly.

The process is relatively simple, but the broad maneuvers of malting are aimed at subtle targets. Too long a soak, too long a growth period, or too much heat can undo everything. Malt starts with live grain and ends at just the right moment of growth. The transformation is more nuanced than milling. Milling is a mechanical procedure, and malting is a biological manipulation of grain. Brewers and distillers, and even millers, want malt that is enzymatically active.

As Andrea and Christian familiarized themselves with the process, they gave their malt to craft brewers that they knew, and to others who sought them out. Word was spreading in New England that malt was on its way. Ben Roesch from Wormtown Brewery in Worcester and Jon Cadoux from Peak Organic in Portland found Valley Malt before the malthouse started production in October 2010.

Early on, the couple saw that they couldn't meet demand. Beer is thirsty stuff. Each gallon drinks up two or more pounds of malt, depending on what kind of beer is being made. The number of barrels in a brewing system is how breweries are described; barrels are 31 gallons, so a 10-barrel craft brewery brews batches of 310 gallons. Valley Malt's original system could make 1 ton of malt each week. A single batch of beer could swallow 700 to 1,000 pounds of malt, or half a week's production.

But tank size wasn't the biggest obstacle. Getting any grains, let alone good ones, proved a challenge. If you want to start making small-batch ketchup, finding decent tomatoes is simple, but grains are an uncommon crop in the Northeast. Corn and soy are the closest field crops grown in quantity. These and any small grains that are grown usually go to animal feed. Brewers and distillers are much fussier customers than cows.

The difference between feed- and food-grade grains is more than two letters. Animals can tolerate a range of protein levels and don't mind grains that might have started to sprout in the field. Once pre-harvest germination has started, however, the game is just about over for malting. Pre-harvest sprouting is not good for wheat and milling, either, but since grinding doesn't rely as much as malting on the internal chemistry of grains, there's more room to use crops with less-than-ideal falling numbers. (A grain's falling number measures how much germination has occurred in the field.) Oddly enough, millers can make some adjustments for falling numbers in flour with, lo and behold, malted barley.

Malting demands very high-quality grains. The grains need to be alive and ready to sprout. One of the first things Andrea tests when she gets a sample is germination. Counting one hundred seeds into a petri dish, she adds 4 milliliters of water and checks growth at twenty-four, forty-eight, and seventy-two hours. One of the factors that can interfere with germination is a farmer drying the crop at too high a heat. Grains have to be stored at 13 or 14 percent moisture, but often have to come off the field in the Northeast at higher levels. Drying grains to store for animal feed or some milling operations is not as touchy as drying for malting; the heat has to be low and slow, just like the type of heat used for kilning or drying.

And that is just one quality consideration. The grains need to be free from weed seeds, fungal diseases, mold, and vomitoxin. Malt has the same 1 ppm limit for vomitoxin or DON as flour does. (Animal feed can be made from grains at 2 ppm.) FDA regulations say that these malting grains must be food-grade, but in practice malting grains have to hit performance measures for seed quality production, not just human consumption.

As they began malting, Andrea and Christian cast a wide net, seeking farmers. Using grains from nearby was important; they wanted their processing to be part of the agriculture around them. This wasn't just about what they could do to the grain, but the full cycle of grains, from ground to glass. Andrea earned the nickname Stalker Babe for her dogged pursuit of grain farmers. She had to be tenacious and follow every lead she got. Those local grains they had read about were sewn up in other projects, like milling for Hungry Ghost Bakery. Wheatberry Café and Bakery had its own mill, and ran a grains CSA. Andrea called the Farm Bureau and extension agents; she went to conferences. This is how they met Thor Oechsner of Oechsner

Farms, who became their friend and ally, selling them grain and helping them start farming.

As Andrea and Christian found farmers, trying to get them to harvest and store at seed-quality levels was tough. Needing to stock the malthouse, they went one step backward and geared up to start growing grains, too, seeking land and equipment.

Their front yard already looked like a miniature experiment station, with rows of barley dominating raised beds. Though decorative and even groovy looking, these tiny patches were very functional, part of the quest to find suitable varieties. Each planting was an inquisition: Will this grow here? How will it malt? Their seed mission was well under way before they considered farming, since finding good varieties was a necessity set by the novelty of their undertaking. Andrea knew the names of barleys that grew in Hadley hundreds of years ago, but who knew what worked in 2010? There is no barley breeding program in the Northeast.

The project of starting a malthouse was much bigger than they could have predicted. They wanted to link beer to agriculture, but they couldn't imagine that they would have to be that agriculture, too. Going ground-to-glass meant finding seeds, kick-starting farming, and figuring out how to malt. It's amazing what they have achieved.

"We know by the smell where we're at," Andrea said the day I met her at the malthouse. She was putting on tall rubber boots, getting ready to hop in the tank and "stir" the malt with a big shovel. "If a bag of grains smells moldy we can't use it. Ninety-five percent of this is having good grain to start. This stuff from Klaas is perfect."

Perfect, she explained, meant that the barley from Klaas Martens and Lakeview Organic Grain had nice low protein levels, and the kernels were plump and uniform in size. If the grains are different sizes, they germinate at different rates, wreaking havoc on the malting process.

I climbed on a ladder and watched her scoop the sprouting grains with a deep shovel, digging down to the floor of the tank and stacking it methodically. She had to be thorough, because the rootlets start hooking together, forming barley bricks and stalling the malting, or causing mold to grow.

"We get attuned to the scents we expect to smell each day," she said. The seeds were on their second day of growth, and as she shoveled I smelled nothing I could name. Porridge? No. Sprouts? Not really. But something was happening, something plant-like and kitchen-y.

When she climbed out of the tank, Andrea gave me a handful of grains to chew. Sprouted barley doesn't taste like much. The hulls are tight to the kernel, and steeping softens the textures. The growing grains taste starchy rather than sweet. Many of the malt flavors come from the Maillard reaction, which happens during kilning and roasting.

"Want to try some malt?" she asked, picking up a woven plastic seed bag that held the grains before they hit the tank. She untied a knot and I took a whiff.

I knew that scent right away. Grape-Nuts! My favorite cereal, made from malted barley and wheat, an accident the Kellogg brothers stumbled upon by forgetting a bowl of batter at their Battle Creek, Michigan, sanitarium, where people went to be cured of nervous and digestive problems. I had made the cereal for my family, using brown sugar to approximate the sweetness of the malt, and whole wheat flour.

"Can I bring some home?" I asked, chewing a few kernels. When she handed me the whole bag, about 5 pounds, I said that was too much.

"You'll use it, right?" she said. Of course I would, but I didn't want to take useful stuff. "Put it in pancakes."

She said the magic word, *pancakes*, and I instantly went from being interested to being fascinated with this food. Looking at the larger world of malted barley helps me see where Valley Malt fits.

Malting makes food, not just beer. Northern Europeans use sprouted grains in dense loaves of bread. Some cultures value malt and malt syrup for nutrition, extending that perceived healthiness to beers. Bakers use malt powder and extracts with deactivated enzymes to add flavor and sweetness, and to assist in browning. Millers add barley malt flour with active enzymes to wheat that tests poorly for falling number; the addition stabilizes the flour's potential to form a gluten matrix.

Brewers and distillers use malt as food for fermentation. The malting process takes advantage of the seed's ability to sprout, letting brewers and distillers sneak in at the right moment to access the starches in the modified grain. Differences in grain types, steep schedules, and kilning styles are where variations on the themes of barley malt occur.

There are base malts, which are used for the bulk sugar feeding in fermentation, and specialty malts, which are used more for color, flavor, and body. About 10 percent of the malt used for an India Pale Ale (IPA) would be specialty malt. The rest might be pale malt or pilsner, two kinds of base malts. Variations in kilning styles and temperature profiles make Munich and Vienna malts. After kilning, roasting allows different expressions from the malted grains. In addition to some of the standard base malts, Valley Malt makes specialties like Cherrywood Smoked Triticale, Danko rye, roasted oats, Munich oats, Warthog wheat, and heirlooms like spelt and Red Fife.

Andrea and Christian are always experimenting: in their words, "pushing the boundaries of what malt can be." They explore their own curiosities, and those of brewers, malting corn, oats, rye, wheat, triticale, and buckwheat. They like to be creative and they have to be creative, because the limited grain supply influences what they use, like wheat.

"We found farmers to grow barley for us, but what was already available was wheat," Andrea said. Warthog, a hard red winter wheat, was already being grown in New York State. That name gave the malt, and the beer that brewers made from it, a distinction commodity grains can't offer.

Regional grain tagging helps brewers and distillers establish their identities, and this has the potential to build a flavor phenomenon, like that associated with the heavy hop character of a West Coast IPA.

"I want to have New England–style beers based on what's being grown locally," Andrea said.

Rye is another good option: grown as a cover crop now, but historically grown as food. The spicy, peppery flavor of rye is getting attention, and malted rye is being used in IPAs and porters, as well as whiskeys. Heirloom corns are of interest to brewers, which is good because New England farmers are growing them. Spelt malts are taking off, too, which is another plus agriculturally, since spelt is less tricky to grow in New York and New England than barley. Many barley crops failed in the Northeast in 2013 and 2014. Being able to malt grains that were not barley was essential for a malthouse dedicated to using what was grown around it.

Further distinctions come from roasting and using a homemade smoker. Valley Malt uses a coffee roaster to make chocolate malt from wheat, which tastes like Cocoa Puffs. Roasted oats are gaining traction with brewers, as are caramel malts. The vessels to roast, smoke, and caramelize malts are too small to make these kinds in quantity, but they keep brewers curious about the malthouse, both professional and homebrewers.

That curiosity has been key in developing relationships with clients and building the business. Commodity farming or malting might keep grain producers and grain users divided, but a new venture like Valley Malt could not exist in a vacuum. Not that I could see Andrea or Christian living or working in isolation. Andrea's very social, and has the ability to draw people into projects. Christian is a friendly guy, too, but his talents lie with objects, seeing where they fit and how they will work together. Andrea has a similar awareness of people and relationships, so she's built that side of the business.

The first year she established a BSA, brewer-supported agriculture, borrowing a name and funding style from the CSA mechanism that helps food enterprises. The BSA linked a brewer to a farmer and a field. The association has grown less exact over time, but the theory and principle remain the same, inviting brewers to help Valley Malt invest all the way back to the land. The spin-off Malt of the Month Club started in 2012: Homebrewers sign up for a subscription delivered in four parts, receiving standard malts and specialties.

These supports function like other CSAs, leveraging financing to help build food and farming outside the dominant and subsidized system. Andrea and Christian chose this marketing tool because they knew they needed solid customers to let them be reliable customers to farmers and develop the supply side of the chain.

I got a good window on the BSA members at dinner one July. About forty people came, bringing growlers and bottles to share in the yard, between the malthouse and the heirloom barley nursery. Under the bulky arms of a beech tree, brewers saluted one another's beers. They talked malt and brewing, and the summer's full roster of special events. Some brewers were headed to Cape Cod the following day for a beer train that one brewer had organized. I asked people what they'd brought, and what they thought of Valley Malt.

"We now have local malt, not only local grains, but malt that meets our specifications," said Chris Loring from Notch Brewing. "Brewers are spoiled because they make our job so easy."

The beer he brought, a Berlinerweisse, didn't sound simple. The brew uses an undermodified or chit wheat, which isn't available in the United States, though you can get some undermodified malt from Germany.

"I knew what I wanted and Andrea thought out the process," he said. "The undermodified malt would create head retention and a kind of doughy, wheaty flavor. We've done a lot of different things with Valley Malt, but I don't think we've done anything as unavailable as chit wheat."

The grains were all grown in Massachusetts, and the formula broke down like this: 10 percent chit wheat, 55 percent wheat, and 45 percent Pils barley malt. Chris makes certain styles of beer year-round, and does a lot of one-offs and seasonal beers. He doesn't have his own brewery, but he brews 4,000 barrels a year at a couple of locations: Mercury Brewing in Ipswich, Massachusetts, and Two Roads Brewing in Connecticut.

Chris is tall and he was standing a little above me on a slope. Figuratively, I felt downhill as well because my knowledge of beer is limited. When another tall man came over, I felt really downhill. Here were two towers of New England craft beer.

Chris introduced me to Ben Roesch from Wormtown Brewery and then left, letting me interrogate him. Knowing he'd been working with Valley Malt since before they opened, I had a lot of questions. I started with asking how he used the BSA.

"Sometimes I'll say what do you have, and I'll build a beer around that, or I'll ask for something special," Ben said of the BSA. Then he described some Wormtown beers. Hopulence and Blonde Cougar all have some Valley Malt in the mix, and the brewery's Mass Whole series uses all Massachusetts-grown barley, wheat, and hops. He poured me his story nicely, without arrogance or reserve. While clearly a champion of Valley Malt, he wasn't gushing praise, just saying how he valued these people and what they do.

"Andrea doesn't have the ability to blend giant batches. She's more like a small brewery. We change as we go along. The malt's not always the same, and that's great," Ben said, naming a limit as an advantage. Other pluses he listed included the different types of malt Valley Malt makes, and the fact that much of it is organic. He also values that the malt is local, and his buying it puts dollars into a business that puts dollars into the farm community.

"Local beer could become a commodity. It becomes a commodity when there's no connection to how it was made," said Ben. "I like knowing Andrea and Christian and how and why they do things. They know how and why we're doing things. There's a lot more to grains than making beer."

Ben didn't know that this thread of visibility along the food chain is what keeps me moving and meeting people in grains. I want to see how everyone sees one another, and how their livelihoods stitch together, even in businesses based hundreds of miles apart. Behind the malthouse, where three shiny silos glowed silvery and pink in that magical light that is a runway for sunset, Ben sold me a ticket to his brewery. Later that month, I'd be heading up to Maine, and I asked if I could come visit. Anytime, he said.

Valley Malt's history is tied to Wormtown. Early in 2010, Ben found them when he was looking for local malt, and Andrea and Christian led him to a farmer who had some rye and wheat. They couldn't malt the grains, but they had a seed cleaner, and the three of them got to know one another over beers, running the rye and wheat through the machine. Once the 1-ton system was operational, Wormtown used the first batch of malt.

The three of them made the beer together, one in a line of what Wormtown called Mass Whole beers, featuring Massachusetts hops, grains, and other ingredients. For malt, that meant products from Valley Malt. Remember that this was New England's first malthouse in quite some time, and a very valuable resource to the brewer. Ben wanted to know what he could do to support the business. Specialty beers would be good, Andrea said, but steady purchasing would do more.

Wormtown puts a bag of Valley Malt in every batch of beer, and many more in its Mass Whole series. This type of buying has been key to Valley Malt's success. Peak Organic, a Maine craft brewery that is building organic farming in the Northeast by using local ingredients, is Valley Malt's biggest customer. The money matters in these relationships, but the understanding that the money represents is just as important. Working on the regional level, brewers have to know the work of maltsters and farmers. Maltsters have to know what both the brewers and the farmers need. Everyone's livelihoods are interdependent.

"Peak and Wormtown, Jack's Abby and Throwback Brewery, and Breucklyn Distilling, all these people are working with us at the scale we're at, and where we're at," said Andrea when I asked her about what made the malthouse work. "It is really helpful for people to not just treat you like you're a vendor, but to actually care about what's going on. And be willing to put up with all the issues that we have dealing with the inconsistencies of the grain supply, and starting off as a new business."

Brewers' support of Valley Malt has been strong, and the exchange is fairly straightforward. The people in Hadley make malt, and brewers buy it. Negotiating relationships with farmers has been less direct, because the raw product they need is not readily available. Valley Malt has experimented with ways to encourage farmers to start growing malting grains. They've made different arrangements, buying seed, or trying to help farmers get land. But they are not independently wealthy and have to keep investing in their own business. Mostly, they've relied on farmers like Thor Oechsner and Klaas Martens, who were willing to take risks on a newcomer because they value new markets for grains.

"Klaas Martens was willing to grow malting barley for us and willing to stick with it, which he didn't have to do," Andrea said. Lakeview Organic has plenty of places to sell grain, but Klaas was willing to work with a small producer. Valley Malt needed grains in one-ton totes, which was extra work, but he and his son Peter did it. "If it weren't for Klaas and Peter, we wouldn't be where we are. Same with Thor. If it weren't for Thor we wouldn't have gotten a consistent or affordable source of wheat that is of the quality that we needed."

Thor also helped them with farming, offering advice on equipment purchases and usage. He looked at land they found to plant, and kept their needs in mind at auctions when he was hunting for equipment himself. Now Andrea and Christian farm about a third of the grains they use. The acres they plant should cover half of what they can malt, but as beginning farmers the learning curve is steep. Plus the weather takes its toll on farmers of any experience, and the past few years have been tough ones for barley.

Over the first five years of Valley Malt, Andrea and Christian kept gathering knowledge and steam. They taught themselves how to malt, and how to source quality grains. They built a malt system and quickly outgrew it.

In 2012, they upgraded the malthouse. They bought the vessels from Malterie Frontenac's original 4-ton malting system and customized the system to suit their space. The four vats are linked to controls. A laptop sits at one end of the malthouse, moving lines on a series of graphs showing temperature, moisture level, and other conditions in the tanks. By computer or phone, Andrea and Christian can monitor what's happening in the grain bed and make decisions.

That same year, they toured floor maltings in the UK, thanks to a grant from the American Distilling Institute. The terms of the grant required that they give a presentation at the organization's conference. The following year Andrea and Christian put what they'd seen to use, building a floor malting area upstairs on the other side of the old potato barn. They bought the barn, and the house adjacent to it. Along the way, they discovered their property overlooks the site where Hadley maltster Andrew Warner's malthouse sat in the seventeenth century. They live where they work, and they are in the middle of malt, looking forward and back.

The Craft Maltsters Guild is an effort to stretch forward the reach of their work. Just as Christian can see how to make metal serve malt, Andrea saw the need for a social structure to distribute the collective knowledge of Valley Malt and other pioneers in craft malting.

Andrea is the president, and the guild is a trade group to share information about equipment and malting. Members have access to a forum where they can consult one another about the technicalities of malting, but knowledge is only one of the resources available. This is a space that is setting standards for the emerging industry. For instance, founding members agreed that grains had to be sourced from the region of the malthouse.

Guild members are also trying to educate themselves and the public about craft malt, and what makes it great. There's an awareness in craft brewing that the entire industry is only as good as each pint; in do-it-yourself endeavors like craft brewing, and now malting, excitement can overrule expertise and execution. The guild seeks to stem this tendency. Just because small-scale malting is starting from scratch doesn't mean novelty should set the tone.

Any one of these projects would be enough for two people, but there is an expansiveness to their busy lives. A sense that more means more, not less. Andrea left her job as a social worker in 2011, and Christian stopped working outside the malthouse in early 2014. All along, they have had a full complement of grandparents nearby to help. The pace of life is a little zooey, but friendly and fun.

Getting into malting has had unintended consequences. A single business, and the series of questions that needed material answers, put grains in the ground and bags of malt in brewers' hands. The beer helps people drink the land. Helping advance the grain system in the Northeast was not their plan, but it is a result. I love learning how Valley Malt evolved, and seeing how change happens in agriculture.

As Andrea tells the story of Valley Malt in different settings, the word *conversation* keeps coming up. The first conversation she had with Jon Cadoux from Peak Organic, on the phone in the parking lot before that NGGA conference, when they were full of information and excitement and only barely scratching the surface of malt. The first chats she and Christian had with Ben from Wormtown, in the driveway on a summer evening, drinking beers and cleaning wheat and rye. The conversation Ben had with Matt Steinberg at Amherst's High Horse Brewery, which led him to Hadley. The conversations at Wormtown the day that Ben, Andrea, and Christian made beer from the first big batch of malt. The conversations with Thor Oechsner about farming and tractors and implements. The talks about transparency in pricing to ensure that everyone in the supply chain gets a fair price.

We discount the power of words, but we should not. Without putting their wonder about malt into words, and asking what new arrangements might be possible, Valley Malt and its ripples back to the land and forward to the brewer would not have happened. Words are powerful, and attached to actions that build relationships. Conversations with suppliers and customers are critical to any business, but established industries are walking a well-worn path. This new venture is cutting a new trail. In the rain. In the dark. You have to keep talking about what you are doing, and what you need, to the people traveling with you. How's the grain? How's the malt? How's the beer?

Valley Malt is making opportunities for more of these conversations to happen. The first Farmer Brewer Winter Weekend was held in January 2012. Andrea brought together experts on grain farming, malting, and brewing for an immersion in malt and the professions on either side of the malthouse. In 2015, the event drew more than a hundred people to Amherst College for intense lessons on the science of malt and the science of farming. Ben Roesch from Wormtown gave a thorough tutorial on starting a brewery. The Saturday-night supper, like any good beer occasion, got a little feel-good and love-festy as people toasted each other and the work they do. But such puffing up is useful. Inside the congratulations there was the camaraderie of shared work. People were cheering one another on, and businesses were growing. When people left the event, they carried hope of the best kind, hope bolstered by information on how to knit something they love, beer, back to the land, and to people.

CHAPTER TWELVE

BEER GROWS GRAINS

*B*eer fans, and many other people, believe that beer is responsible for the agricultural revolution, not bread. People started growing grains to make beer, not food. Flour is my true north, so I align myself with the bread story, but all around me I see evidence that beer is indeed putting grains in the ground.

New England is a good place to observe this in action thanks to Valley Malt and the craft breweries that put local ingredients front and center.

Cambridge Brewing Company's brewmaster Will Meyers is a sort of god-father for craft brewing in New England. His skills and instincts are evident in the brewers who have worked with him, and one of those signature traits is the use of local ingredients.

The habit came from reading old beer recipes, which got him intrigued with special ingredients. Pumpkin ale may seem a bad attempt to squeeze pie in a bottle, but fermented pumpkin beverages were actually a necessity in colonial New England, where barley and malt was often in short supply.

The stories of beer intrigued him, too. Will was immersing himself in work that had little to do with the narrative of mainstream beer, but mainstream beer was the frame of reference for the audience craft beer was attracting. He was struck by the differences in the way that wine and beer producers presented themselves. The wine industry romanticized the idea of a man walking the vineyards, picking grapes, even though most winemakers do not grow their own grapes.

"I could see how industrial breweries had failed in maintaining that connection to agriculture," said Will. The closest that beer ads came to nature was zeroing in on the skin around a bikini, or dipping a six-pack in an icy-looking stream. Craft brewers restored the idea of individuals toiling to make beer by hand, yet they were forced to use the same ingredients as industry giants. Where, he wondered, was beer's terroir?

Long before local malt became available, Will sourced local pumpkins for Great Pumpkin Ale. He brought people to walk the Minuteman Trail hand-harvesting wild hops, and foraging for heather to make CBC's Heather Ale.

Local ingredients feature in the beers, and on the brewpub's menu, too. They serve hamburgers made from grassfed Massachusetts beef, and vegetables and cheeses from nearby dairies. On their website, the background is a picture of a field of grain, and Will is glad to have access to real grains from Valley Malt and to explore Massachusetts terroir.

When Ben Roesch first worked at Cambridge Brewing, he already had a strong connection to the land. Ben, the founding brewer at Wormtown Brewery, was studying forestry at UMass–Amherst when he started home brewing. His classes took him out to farms, and he brought home ingredients to use in his beers. After college he worked at Land's Sake, a land conservancy and farm outside Boston. He was the forester and ran a sugar shack. When his work started to shift from land to beer, he brought the land along, or at least its products. Will Meyers started buying organic sugar pumpkins for CBC's Great Pumpkin Ale from the farm at Land's Sake right away.

Ben worked part-time at Cambridge Brewing, and part time at Wachusett, another craft brewery, for two years. Next he was head brewer and distiller at Nashoba Valley Spirits, where he had access to an orchard and could bring the land indoors, incorporating seasonal ingredients into all the beers and spirits.

He really got the itch to make his own place by designing the brew system at Honest Town Brewing Company in Southbridge. Ben had always dreamt of opening a brewery in his hometown, Worcester, a rough-around-the-edges relic of New England's factory era. Worcester is the second-largest city in New England and heavily populated with colleges, so it seemed like someone else would beat him to it. A brewpub in the 1990s didn't last, and perhaps its failure kept other people from taking the leap.

By 2009, restaurant owner Tom Oliveri was ready. Making beer looked like it might provide a steadier income than the ice cream parlor he'd built at his restaurant, Peppercorns. Tom bought the beer Ben made at Honest Town, and the two men started talking about turning the small space into a brewery.

The brewery opened on St. Patrick's Day, 2010. Peppercorns and Wormtown Brewery share a parking lot and a roof. The restaurant has a nice bar and serves as a tasting room. People can tour the truly microbrewery, which is less than 1,000 square feet, in a few seconds. All of its pieces are stacked and scrunched together. Brewers look like sticks, working between the stout and shiny steel tanks for mashing and fermenting. Everything feels too tight to make anything, but the 10-barrel brewhouse makes 3,000 barrels of beer each year. A desk passes for an office, sandwiched between rows of kegs, most of them the smaller, slender style that dispenses soda syrup underneath bars.

Outside, a storage container holds grains and a mill. An auger moves cracked grain from the mill into the mash tank. Wormtown outgrew the old ice cream parlor almost immediately, and expanded its original system within a year. Four years after opening, they started to build a whole new brewery.

This new operation, like the old one, will be doubly steeped in the concept and practice of local brewing. Wormtown boasts that it put "A Piece of Mass in Every Glass," using at least one Massachusetts ingredient in every batch of beer. Usually that means a bag of malt from Valley Malt.

Approaching its fifth anniversary, Wormtown remains the city's only brewery. The beer has an identity that is, like craft beers in general, rooted in place, and in the brewers' skills. This is Worcester's beer, and it is known in and out of town for consistency and quality. At the U.S. Craft Beer Open in 2014, Wormtown walked away with Best in Show honors.

In the new location, the sense of place and attention to quality will prevail, said Tom and Ben's new business partner, David Fields. He has a long link to the brewery. Consolidated Beverages, an Anheuser-Busch distributorship he recently sold, was Wormtown's first distributor. David is also from Worcester, and he ties the brewery's identity to a sense of community shaped by the city's practical housing style.

"We've got a little bit of an edgy three-decker attitude," he said. "A three-decker is where you live. If you look at Worcester and other cities, Lowell and Lawrence, we grew up in three-decker houses. That's the environment we're all raised in; it made for good neighborhoods and it made for a good city and it made for a good strong blue-collar work ethic."

Their flagship brew in and around Worcester, Seven Hills Pale Ale, is emblematic of the brewery's dual strongholds, with one foot in the selective realm of craft beer and the other firmly planted with local Janes and Joes. David describes Seven Hills as a gateway craft beer, something that someone new to craft brewing can appreciate just as much as someone more acquainted with the bigger tastes of craft beer.

Beyond their home turf, Be Hoppy IPA sells much better than Seven Hills, even in areas that have a similar demographic, like Springfield. The discrepancy baffles David. He, Tom, and Ben try to get a bead on what sells and why. What motivates people to choose a Massachusetts brand over a national brand of equivalent quality? The choice is probably tied to the idea of a specific place. A locale can sell a beer, but does localness sell beer, too? Despite Wormtown's use of local ingredients, this is not easy to answer.

The lore of craft beer began with people. Brewers who could make beers with atypical, off-the-grid tastes received praise and notoriety. A decade ago there were a thousand craft breweries in America and now there are more than three thousand. As the industry has grown, the focus has shifted from using novel ingredients from wherever to using hops and malts with strong identities. Ben's inclination to make beers from local materials has helped distinguish Wormtown, but this isn't market positioning. Ben's disposition led the brewery to highlight local products. A guy who studies forestry and works for a land trust has some gut feelings for agriculture and nature.

"A Piece of Mass in Every Glass." The theme came from that beer that Ben made with Andrea and Christian from Valley Malt's first big batch of malt. They were making a Mass Whole beer, the name Ben gave the line of brews he made from Massachusetts products. That day, in the tight quarters of the old ice cream parlor, they discussed how Wormtown could best help the malthouse. If each batch included a bag of Valley Malt, that would help support the fledgling business. As the owner of a new business himself, Ben had an appreciation for the uncertainty of the enterprise.

Plus, Ben wanted to make sure Valley Malt would stick around. This was the only local malt he could get. (There still isn't another malthouse in Massachusetts.) Making a consistent commitment, rather than giving random support through one-off batches, made sense.

Wormtown buried the expense of supporting higher-priced ingredients by leveling the prices they charge for beers. While heavily hopped beers and beers with specialty and local malts cost more to make, the brewery spreads the cost across the boards.

Commodity malt can sell for as little as thirty cents a pound. For a number of reasons, custom regional malts cost more. Small malthouses are aiming for unique products, and buying from farmers whose expenses are not insulated by commodity markets and subsidies. Lacking efficiencies of scale, craft malts can cost a dollar or two a pound.

By selling everything they make at a reasonable price, Wormtown doesn't put a price burden on local.

"There's a small percentage of people willing to pay extra for organic, and there's also a small amount of people willing to pay extra for local," said Ben, speaking of organics and local food in general, not just beer. "We can offer that extra dimension of being local and high quality, and people don't have to make a decision that will cost them extra."

Driving from the original brewery to the new location, you go under the elevated highway, moving from a gritty, lived-in area to one that's much more lively. If the old neighborhood has a gray and dirty face, the new one is clean, a blank slate for a good time. Streetlights and plantings are sparky, and new sidewalks connect a string of places to eat and drink.

The new facility used to be a Buick service center. A viewing and tasting room show off the production area, divided by a big glass wall. There is space for grain storage and cold storage of kegs, and the brewing system is twice the size of what they had. With this buildout, they anticipated extra growth, and now have the capacity to make five times as much beer. They've added a bottling line that can handle 5,000 bottles an hour, a jump from the old device that bottled only 120 an hour. Outside, there will eventually be a silo holding base malt and painted with a big red *W*. The malt inside will come from the Midwest. However, this symbol of farming will be in the city, visible at street level and from the highway. A sign of reverse agricultural drift.

As Wormtown got ready to move into its new location, the brewery hired a new head brewer, Megan Parisi. Her first job in brewing was also with Will Meyers, and she fondly recalls once collecting 11 pounds of heather for a batch of beer. Brewing can be its own world, an intense and complicated flow of time and materials. Using herbs, hops, and malts with bright flavors and traceable roots keeps the process grounded for her. The freshness is an advantage, too, and gives your beer a chance to be more unique.

"If we're all getting our malt from the same malthouses and hops from the same farms, a pretty large percentage of what's taking place in the beer is the same," she said. She's glad to be part of a brewery that wants to explore ways for the beer to express the region.

Of course, Wormtown doesn't use local ingredients exclusively. That would be impossible. There isn't enough grain grown or malt made in New England to meet their needs. They use hops from New England and New York, and elsewhere. Hop Union, a growers' cooperative in Washington State, provides nice options. Megan and Ben went to Hop School there and spent a few days learning more about this part of beer.

The fact that Hop Union is a cooperative resonates for Ben, who likes to keep business relations personal. That's part of the appeal of local ingredients to him. Yes, they are fresher, and he values the environmental stewardship of local production. But most important to him is that people are in the process.

"The local ingredients support local people and their ability to do things that wouldn't be able to be done," he said. "So this supports farmland. This supports agriculture, and people I actually interact with. This is a product made by people I care about."

Being able to see and know the people who are working for beer keeps his livelihood out of the commodity realm. As the brewery grows, he wants to keep the beer handcrafted and not shove people out of the process. The new facility has streamlined operations that make brewing physically easier, but each step still requires people. If brewers were just flipping switches, the product would become a commodity, and the brewery would be just another business. He never wants to lose sight of the human element.

"I'm always going to be talking about the beer and the raw ingredients, not just units in and out," he said. "It would be a lie if I didn't talk about the bottom line. These are people's lives. Without having a business that's financially stable, you can't be responsible to your employees."

This responsibility for the welfare of others travels back to the land, through the choices he makes in sourcing. By proxy, this man who trained to care for trees is still tending the land.

Jon Cadoux has a similar mission of stewardship. Peak Organic Brewing Company has a goal to build organic agriculture in New England.

The Portland, Maine, brewery has its roots in weekend homebrews in the 1990s. Jon and some friends from summer camp would meet in the different New England towns where they lived and make beer. That hobby has now become their careers, thanks to Jon's passion.

"I was a junior in college when I realized organic beer is what's going to happen, and it's going to be sourced locally from New England farms," he said. He was majoring in psychology at Middlebury College in Vermont. Once he had that vision, everything he did was in support of building the brewery, including getting an MBA degree from Harvard.

Organic sourcing in the Northeast was a tall order. Jon had to build supply, courting farmers to grow grains and hops. He began by talking to people at colleges, conferences, cooperative extensions, telling anyone and everyone, "We're Peak, we're here, we're interested." Connecting with farmers requires a constant barrage of inquiries, and Jon never stops getting the word out.

"You have to be extremely proactive to make this all happen," Jon said. "It's so easy to do it the other way. All you have to do is say yes and the stuff shows up at the door."

The brewery opened in 2007, and since that time Peak has nudged along organic production in a serious fashion. From zero choices, Peak now has a steady group of growers, some of whom have contracts that extend forward five or ten years. The timing varies, depending on what growers need to feel secure.

"We're not going anywhere," Jon said he tells people. "We'll buy it from you this year, next year, and the year after that. You'll know where it's going. Peak is going to be here."

Jon has been effective at growing ingredients for the beer movement, broadcasting Peak's presence and intent like a refrain. This insistence is a necessity in farming in general, and especially in asking farmers to switch to organic production. If a grower starts the three-year transition required for organic certification, knowing a customer like Peak is going to be there at the end will help cushion the investment.

One of the reasons so many growers stick with standard options and sell to commodity markets is because they exist, and have existed. Commodity markets might not pay much, but they will be there every year, ready to buy what farmers grow. Peak's record of purchasing from organic farmers now stands to back up Jon's claims, and the company gets leverage from making something people love—good beer. Soliciting farm partners for something like organic dog treats might not have the same appeal.

"Most of the growers we work with love beer, and they love the idea of something they grow going to the store," Jon said.

Farmers' stories and the farm-to-glass ethic come through in the way the company tells its story. This is not through paid advertising, but in social media and via a team of committed salespeople who connect buyers and the public with the farmers who make the beer possible. Online, people can find stories about the people who grow the ingredients, snapshots and written profiles, and interviews. Peak celebrates harvests and producers by hosting events at farms and holding festivals.

Jon found Valley Malt before it was even open, a sign of the diligence of his hunt for local organic ingredients. He considers Valley Malt a great boon to Peak, and its malt is in every beer Peak brews. The malthouse is helpful in building relationships between Peak and farmers, too. Sometimes Andrea Stanley will find farmers and send them to Peak, or sometimes Jon will find growers and direct them to Andrea.

Andrea values the partnership and the work that Peak does for organic farming. She traces about 600 acres of organic grains to the company.

"That's good for growing," she said. "That's good for us and our business, because they are a steady customer. But they are not just a customer. Peak is a customer that's listening and responding."

Listening and flexibility matter because Valley Malt is using material that's more fickle than predictable. Any crop is subject to variable expressions thanks to different soils and growing seasons. A crop that hasn't been grown in New England in over a century, like barley, is going to perform more erratically. As growing barley for malting in the region becomes more commonplace, some of these irregularities will be ironed out.

One area that needs ironing is seed varieties, but there is no barley breeding program in the Northeast. The University of Vermont and Cornell have been trialing barleys on their own as they seek funding for formal evaluations. Private industry is involved in the quest, too.

KWS, a German seed company with more than 150 years in the industry, launched its American cereals branch in Illinois in 2011. Cornell, Penn State, and Ohio State have been testing KWS barley varieties. While the performance results won't be available for another year or so, sample amounts of a couple of KWS barley types are for sale through a commercial seed company. Farmers can begin to grow them and see how they work.

New York growers have a great curiosity about barley, inspired by the Farm Brewery Law, which mandates the use of in-state ingredients. The law eases licensing requirements for breweries and has begun a swarm of malt-related business activity. Clusters of craft malthouses are popping up around Rochester and Buffalo, regions where grain farming is more common than other parts of New York. Bill Verbeten, Cornell Cooperative Extension agronomist, and his colleagues are trying to help farmers figure out the puzzle of how to grow malting barley. Bill works with all field crops but has taken a special interest in barley because of demand. The State of New York is now financing barley variety research at Cornell to back up the need that the law created.

The law is a potent economic driver; it has generated so much interest that the state liquor authority set up an office dedicated to helping people apply for a Farm Brewery or Farm Cidery license. Both of these went into effect in 2013, inspired by the success of the earlier Farm Distillery Law and its parent, Farm Winery licensing. Wineries are the only ones that must be located on a vineyard; the rest of the state licenses are tied to usage of farm products, not production. Farm Winery legislation passed in 1976, fostering both the New York wine industry and a corresponding agritourism business. Cornell research led to hybrid grape varieties geared to our region, and vintners also found their own way to grape varieties suited to the many microclimates in the state, sometimes turning to public breeding programs in other states for guidance. This kind of broad search is under way for farmers who want to grow barley, too.

Cornell is newly involved in malting barley research, but North Dakota, Minnesota, Oregon, and other states near the country's main barley-growing regions have long-established breeding programs for the crop. As is the case with wheat, public barley breeding programs focus on identifying and creating varieties suited to industry standards. Part of the land grant mission is to conduct research on plants of economic importance; if the main buyers are Big Malt and Big Beer, that is going to determine the scope of a breeding program's investigations. Funding sources also direct those arrows, but the terrain is starting to shift as craft brewers take an interest in malts.

"Brewers were working through so many options in terms of hops and yeast and beer styles," said Pat Hayes, barley breeder at Oregon State University. "The barley piece was one constant."

The variable of barley and its contributions to flavor are now piquing brewers' curiosity, especially in craft brewing.

Pat Hayes has been working with barley at OSU Corvallis for nearly thirty years. He calls barley "the fruit fly of crop plants." By that he means, compared with wheat, the genetics of barley are much simpler, so it is easier to analyze in the breeding process. The OSU program has focused on trying to understand the genetics of what makes barley useful to the malting and brewing industries. The explorations have stretched into food, too. Hulless barleys are drawing interest because of their high fiber, lower gluten content, and relatively low glycemic index. Deeply toned purple and black barleys are high in antioxidants and have flavors that chefs love.

Until recently, malting barley research efforts served large brewing and malting companies in the United States. Pat Hayes and his colleagues looked at things like endosperm characteristics for starch synthesis, which affects how malting and then brewing can access the grains' desired starches. Researchers explored winter growth habit and temperature tolerance to find fall-planted barleys that could better suit the changing climate. Most of the studies looked at six-row barley, which has been most commonly grown and used in America. (Six-row barleys have six rows of barley kernels in each head; two-row has two rows of kernels on each head.) Much like the metric system, most of the world uses two-row barley. Now more American growers and breweries are using two-row as well. The rest of the ground in barley research is shifting, too, as the world of brewing changes.

Hayes describes the relationships with Anheuser-Busch and AMBA, the American Malting Barley Association, as wonderful, because they provide funding with no restrictions. The companies don't own what is created, and the researchers leverage that basic funding to get grants from the National Science Foundation and USDA.

There is commonality to what a mainstream brewer and a craft brewer want. Both want high wort extract, which is the amount of soluble carbo-hydrate in the malt that will feed the yeast in the wort. Plump kernels are a must. Two-row barley is preferred by both types of brewers—the dividing line is around adjuncts. Craft brewers pride themselves in brewing with an all-grain schedule, but mainstream brewers often add corn, rice, or sugar as supplemental starches in the brewing process. In those cases, the barley is the enzyme source to break down the additional starch, which necessitates a higher level of enzyme activity in the grain. Craft brewers don't need such elevated enzyme activity, since each malted kernel is responsible for converting its own starch, not utilizing the starches in other ingredients.

Over the last century of barley breeding, programs targeted agronomic goals for farmers, and quality factors other than flavor for malting and brewing. The idea was to get barley that did its job well. Introducing flavors was just not on the radar, and as the research focus shifts, what the grain contributes to flavor remains somewhat of a mystery.

Sierra Nevada and Rogue Brewing have been leaders in exploring the frontiers of malt, bringing farms and malting into their operations. At the research level, exploration of that frontier, and funding for it, is increasing.

The American Malting Barley Association acts as an intermediary between industry and research. AMBA began in 1938, following the repeal of Prohibition. There wasn't good information on barleys at the time, and the group was established to fill the void. AMBA is a membership organization, collecting dues from brewing and malting companies to fund public-sector barley research and develop new varieties.

AMBA collaborates with barley breeding programs to develop competitive barley types, surveying its membership on desired specifications and trialing university-bred varieties. AMBA guidelines set the analytical and performance targets for barley breeders. A few years ago, AMBA turned its inquiry toward breeders, too, looking for input on unusual barley characteristics to see if its membership had interest in exploring the contributions barley makes to beer flavors. When AMBA presented this data to members, a definite interest was voiced.

AMBA is a unique trade organization because it represents both artisan and industrial scales of beer and malt. Two levels, and fees based on the volume of sales, divide the groups into Regular and Associate members. That craft and mainstream brewers, who often don't have a lot of love for each other, in this case share a common alliance is intriguing and useful.

New barleys matter to all breweries, from micros to monoliths. The definition of a craft brewery is murky. Today there are big breweries trying to look little, and small breweries that have ballooned into big companies after expanding the meaning and taste of beer.

Whatever the definition, no one in the brewing industry can ignore the impact of craft beer. The percentage of sales for the craft sector, relative to production volumes, is too high. Craft brewing is capturing an intimidating slice of the market, so large-scale breweries are taking cues from craft traits. Enter the thunder of the heavily hopped India Pale Ale (IPA) style, now made by almost everyone under the sun, from small breweries to big brands hiding behind crafty labels. Craft malts are another way brewers can distinguish themselves in a crowded marketplace.

Since AMBA has members across the beer divide, the organization is now also driving research suited to craft malt, craft beer, craft distilling, and their counterparts. Toward this end, the organization added a new layer of targets: Barley breeders now can reach for technical specifications geared to six-row, adjunct two-row, and all-malt two-row barleys. Though this sounds like it could just make flavors the equivalent of the color brown, rather than finding a series of distinct colors (that is, flavors), the craft industry, in both malt and beer, is still identifying its needs in terms of perceptual and flavor issues, as well as nuts-and-bolts performance issues. So mud may be a good enough color to find at this stage of the game.

"The challenge is trying to get a grip on what on earth flavor in barley could be. A whole lot of time and work has gone into flavor and wine," said Pat, but the work in barley is just beginning. The interrogation took them all the way back to very wild and primitive barleys, trying to find and name the beer flavors that come from barley.

Novelty barley research is being funded through alternative streams, such as the Flavor Six Pack. This group of craft brewers is directly expressing a financial interest in new malts. By pooling money to fund a graduate student at OSU Corvallis, they're pushing the next horizon in beer styles. Breweries include Bell's, Firestone-Walker, New Glarus, Russian River, Sierra Nevada, and Summit. Rahr Malting, Briess Malting, and the CMBTC (Canadian Malting Barley Technical Commission) have provided malting for the trials because OSU's pilot-scale malt equipment doesn't yet have a laboratory. In the first round of trials, beers will be brewed at New Glarus and Sierra Nevada. The long-term plan is to house fermentation facilities on campus for brewing analysis of the malts.

Teasing out specific flavor components in beer, and how they travel from seeds and through the malting process, is complex. Different soils and seasonal growing conditions can change a grain's composition, so trial varieties planted at multiple locations try to equalize the differences. Interactions among barley, yeast, and hops are controllable only to a certain extent, which makes identifying flavors both exciting and humbling. This is the case with bread, too, where the flavors of fermentation influence the tastes of wheat.

The collaboration has brought to the fore some barleys of interest, such as Full Pint. This variety didn't meet AMBA guidelines when it was tested a decade ago, but researchers at OSU thought it worthy of a closer look. Oregon farmer Seth Klann got word of the variety and grew out a crop. Once malted, a beer swoon followed. Sierra Nevada liked the way it brewed, and the farmer and his family decided to build a malting operation, Mecca Grade Estate Malt, around Full Pint.

Mecca Grade is unique for a few reasons. Not many places grow their own grains, let alone build a malthouse around a single type of barley. The size of the facility is different, too. While small malthouses in the Northeast make 1-ton batches, that's the size of Mecca Grade's mini floor-malting systems. There will be three of these, and two of the larger malting systems. These will make 10 tons at a time and use a new technology they've developed, mechanical floor malting. Neither giant nor miniature, Mecca Grade is helping establish another scale. Skagit Valley Malting in Washington State is also working at that bigger-than-little level, and so is Colorado Malting, one of the oldest craft malting facilities in the country.

The farmers who started Colorado Malting came from a family that had grown commodity barley for fifty years. They began malting as many malthouses do in New York, converting an old dairy tank into a malt vessel in 2008. Over time they've expanded their equipment; they now malt 1.2 million pounds of barley a year.

As regional grain systems evolve, the question of scale is tricky. Farmers can't plant for micro-facilities that might only use 100 acres of barley each year. That amount is hardly enough to have a commercial grower drag out the combine. Yet finding money to finance a facility that would demand grains at a rate that would be worthwhile to farmers is tough. So maltsters are starting at the scale they can afford and scrambling to find barley. Farmers are hanging back in some areas, looking for varieties that will work in their weather and soils. But they are also waiting for customers who can support their migration to another crop. And if growers can't transition, how can breweries like Wormtown ever get enough regionally grown grain, let alone have it malted by a facility that no one can afford to build? This is a piggybacking stack of ifs that makes the idea of a regional grain system seem impossible.

Having commercial interests in multiple uses for grain can help sort out the issue of scale, at least for growers. This is the case in North Carolina, where beer, malt, and bread are growing grains, without a doubt.

Brian Simpson and Brent Manning started Riverbend Malt House about the same time as Valley Malt, in 2010. Riverbend evolved vicariously from work in environmental consulting. The friends wanted to create a career centered on sustainability. Brian had run a biodiesel cooperative, but they wanted to do something that also contributed directly to the community.

"We wanted to attach ourselves to the ag economy," Brian said. They knew what grew around them, but wondered where the grain came from in Asheville's craft beer explosion. "North Carolina has all these grains, mostly wheat and rye. Why can't we utilize it to make beer?"

Brent was homebrewing at the time, so Brian started floor malting in his basement, getting research grains from North Carolina State. Using a steel tub as a steep tank, and a tile floor for germination, Brian recommissioned an electric smoker as a kiln. Each round made 12 to 15 pounds of malt, enough for a 5-gallon batch of beer.

The single-tank system more common to craft malting had no appeal for them. They liked the traditions of English and Scottish floor malting, and also the physical connection of actually turning the malt by hand. Being able to watch the grains sprout was helpful as they learned the process, and that involvement is still worthwhile now that they know more. He and Brent can observe conditions better than they could in a tank. They got more training at the CMBTC.

"We called up and said, 'Would you teach us to do this?'" Brian said. The western Canada research facility serves large-scale brewers and growers. They'd never had such a request, but Brent and Brian spent almost a month with them, learning in classrooms on a small-scale malt system. Now the CMBTC has programs geared to craft malt, offering brief workshops to teach the basic science of malting.

Riverbend started with capacity to make a ton at a time, moving grain by hand between steep tank, floor malt, and kiln. They made a vegetable distribution center work for malting, converting a walk-in cooler into the germination floor. Probably the first day they made malt, they saw that their size was insufficient. But they couldn't have built any bigger. Colorado Malting was a solid but isolated example of craft malting.

"You've got a niche product. You don't know if anybody's going to buy it, and you're using somebody else's money to buy grain," he said. "That's already scary. We had no proof of concept. We thought we'd just do a little system and see who buys it."

They initially had 2,500 square feet, and three years later began to expand on the other side of the building, into 9,000 square feet. Now they have augers to move the grain around for some parts of the process, and plenty of storage for raw and converted grains. Eighty percent of what they malt is barley, and the rest is wheat and rye.

The feedback the maltsters get from floor malting helps them manage the malting process. Grains they get might not have a lot of analysis to suggest handling methods, but watching the grain every few hours, rolling around 4 to 5 tons, allows them to respond to what's happening.

"I can keep more heat in the grain for longer" if it needs it, Brian said. "I can thicken the bed or thin it, and speed up and slow down the germ process."

Other parts of the system have room for adjustment, too, like the temperature and time on the steep tank, and the same for kilning.

At first, Riverbend had to seek out growers and give them seed. They had some issues with barley quality, but the quality improves every year as growers understand what the crop needs. Though mycotoxins like fusarium can also be a problem in this region of the country, it hasn't been hit by any failures due to weather or disease. Brian and Brent work with about five growers, and, to accommodate their expansion, are increasing that to eight. This is straightforward because, after four years, they're now getting calls from farmers.

They work with North Carolina State for variety information, and get some support for malt analysis from Virginia Tech. The business is an AMBA member, though at first they didn't see the benefit of joining what they saw as a corporate entity. The region-specific research that AMBA provides is good for the whole industry moving forward.

The benefits the Craft Maltsters Guild offers were more obvious, and Brent and Brian were involved when it was just a Google group of maltsters swapping information. Brent is currently the guild's treasurer. The connections and resources the guild provides start-ups are valuable, and both Brent and Brian believe that having a mechanism to set standards for the emerging industry is critical.

More locally, interconnected grain projects have bolstered the malthouse. Jennifer Lapidus, the founder of Carolina Ground, started a network of non-commodity organic grain growers in response to soaring flour prices in 2008. The North Carolina Organic Bread Flour Project united bakers and growers with money from community organizations and tobacco groups. The region abounds with miller-bakers and baker all-stars, and growers that fit their demand for high-quality grains. Jennifer Lapidus has helped Riverbend with connections to farmers and logistics, and the malthouse can sometimes buy wheat that doesn't meet milling specifications.

What Brian likes most about this, his third career—he ran a hotel on a small island for a time—is sitting in the back corner, watching people drink beer with malt he's made.

"There's something about our connection with our farmers, and seeing how it actually affects our community. We're taking the global commodity market out of the picture," he said. "Nobody knows how a beer is made. Now they know it comes out of the equipment in the brewhouses. They don't know about the malt, but they will. Then the question will be about the farms."

The snugger the supply chain gets, the better. People will keep seeing people and follow that feel-good, ground-to-glass theory back to the land. The last back-to-the-land movement, the one that sent people to the woods in the 1960s and 1970s, was all about isolation, turning away from society. This one is about connection, and turning to society. That first impulse was, and this second one is, subversive.

CHAPTER THIRTEEN

THE NEXT BREAD

J have met the next bread.

I'm more than a little susceptible to hypnosis by wheat, but high on a hill in Vermont at Elmore Mountain Bread, I've found a future for flour that might last.

Blair Marvin and Andrew Heyn bake sourdough bread in a wood-fired brick oven, which is standard operating procedure for artisan bread. However, they also mill their own flour. More bakers these days are adding stone mills to their kitchens because the process allows them to use more whole-grain flours and experiment with flavors.

Remember when bread was benign, before the gluten-free craze? Some people are finding their way back to wheat through small-scale bakeries and long fermentations. The next road on the path back to bread may be house-milled grains.

Grain kernels have three parts: bran, endosperm, and germ. Most of the oils are in the germ and the bran, which also hold minerals, nutrients, and flavors. Flavor and fat are volatile. Once exposed to air in the milling process, the oils in grains spoil quickly. Bran has other strikes against it, and the biggest is that it interferes with making lofty, airy loaves of bread.

Roller mills, which were adopted in the late 1800s, allow for removal of bran and germ. One advantage of this is shelf stability, and another is making flour that is mostly endosperm, a powerhouse of starch and protein that is great for baking.

Stone milling was the way to make flour for millennia. Now millstones are artifacts of how we used to live, visible mostly at historic sites. However, the technology is having a revival. People have a hunch, backed up by preliminary research, that the practices of artisan bread baking can alleviate health problems attributed to gluten and wheat. Wild yeasts and fresh, stone-ground whole-grain flour are valued for their contributions to flavor and nutrition.

"We want to make the best bread we can, and it's a no-brainer that milling is a part of it," said Blair. We were traveling in a 50-mile radius to mom-and-pop stores, restaurants, and supermarkets. Bread filled the back of the Honda Element, and a cargo box on top. Andrew was driving a similar load in a matching car.

Blair had changed into a skirt for deliveries, tucking her tawny hair in a bun. At a small supermarket, she transferred loaves from a black plastic tote onto the racks, the birds in the tattoo on her arm flitting forward, moving but staying still. Like those birds, inked and trapped on her arm, Blair is simultaneously contained and outgoing, quick to engage with anyone she meets. The exchanges are brief, and often don't get past hello. She doesn't have much of a chance to discuss the radical changes they've made to the bread. A tag clipped to the racks announces they've started milling their own flour. The bread bags have a red stamp of a millstone, wrapped with the words ORGANIC STONE-GROUND WHEAT, MILLED FRESH DAILY.

These notices are small, as is the time that bakers and bread generally have to tell their stories. A beer, by comparison, gets its own space. Bars and brewpubs are places to hang out, and the beer is a conversation piece. Sitting with a friend, you can muse fondly about ingredients and their provenance, and the skills of the brewer. Farmers markets offer more room for people to get to know a loaf, but even then the talk might only be confirmations of suspicions rather than a real dialogue. Generally, we approach beer from another angle, when we are ready to relax.

Blair and Andrew sold at farmers markets for seven years, finally shifting their distribution program after they had their son, Phineas, and needed to carve out time to be a family. Aside from notes on racks and bags, they don't have much access to their audience, or a way to communicate about the four hundred loaves they make three times a week.

The bread has to speak for itself, and it does, fluently articulating the language of artisan baking. These are nice hearth loaves, dark-crusted and dusty with flour, bursting just enough at the cuts to unzip the dough's energy in the oven. Inside, the bread looks earthy and bright. Some loaves are almost orange, like a carrot just pulled from the ground. The baguettes are much more creamy-colored, and their crumb is more open. The perfume of all the loaves is full of well-tended sourdough starters and well-handled flour. The Brewer's Bread smells more sweet than brewy. The Country French, Seven Grain, and other loaves are lightly tangy and sour.

I didn't taste any of these breads before the flour changed, but Elmore Mountain Bread made me forget about my griddle for days. Normally, all I want is pancakes, but with that bread in the house I had no other flour on my mind. After my family and I devoured them, the memories of the breads hung around like ghosts. I kept remembering my visit, trying to understand how simple things add up to a revolution.

Up, up, up a dirt road on the edge of Vermont's Northeast Kingdom, Elmore Mountain Bread is attached to Blair and Andrew's house. Stacks of wood fill a tall shed, same as at any woodsy dwelling. The only clue to the mill is a perpetual snow that dusts the shingles on the garage, below a vent for the exhaust system.

Making flour begins by pouring grain into a hopper above a set of granite stones. The mill, designed and built by Andrew, grinds away while he and Blair work in the bakery. The freshly ground flour flows over to a sifter that shakes off some of the bran. Every half hour, one of the bakers checks on the mill. The process allows them to bake with flour that uses most of the whole grain.

The idea for milling came from a few cues. Elmore Mountain Bread is remote, but a hot spot on the bread circuit. Good bread advances through a network of online and live resources, and peers travel around the world to see one another's setups. Blair and Andrew's equipment is a powerful magnet. Before they built the mill, they built their brick oven. This bakery is a laboratory for hacking traditions on the way to good old-fashioned bread.

"One of our missions has been figuring out how to do things more efficiently, and with less wear and tear on our bodies," Blair explained early one evening in August. She and Andrew shaped pieces of dough into nice round balls. "Our goal is to be able to produce more, higher-quality bread in less time without having employees. But also without compromising the quality."

For instance, bakeries of this scale don't often use mechanical dividers. Dividing the dough can take multiple cuts per loaf, which makes a lot of shock for wrists and forearms to absorb. Instead of using peels to load the oven, Andrew and Blair use a canvas-belted mechanical conveyor. These machines don't compromise the quality of the bread, just limit the repetitive motions that can cause injuries.

While shaping loaves, the couple consulted about who would take the babysitter home. Andrew left the bakery and entered the house, kicking off his flour-coated clogs at the door.

"I was young enough when we got into this that I knew I wanted to protect my body," said Blair. She was twenty-four when they bought the bakery, but doesn't look old enough to have been doing anything for a decade.

Andrew is slightly older. He comes from a family of engineers, so he is inclined to fiddle with objects and systems, and improve the ergonomics of a physically demanding craft.

The next morning, Andrew got up at five to load the oven with wood, lighting a fire directly on the same hearth that would bake the bread. Because the oven is fired three days a week, the residual heat is significant. By seven, when I arrived, the oven was sufficiently heated, and Andrew was scraping ash through its long, narrow mouth. The insulated metal door, about 6 feet long and a foot wide, stood near him, resting against the bricks. Andrew moved the ash outside with a wheelbarrow and swabbed the deck clean with a wet towel. He loaded sheet trays with wet towels to cool the oven. The water had been weighed to calibrate the cooling and bring the temperature down from 750 to 600 degrees Fahrenheit (284 to 227 degrees Celsius), the temperature needed to bake the first breads.

As he worked, I asked questions and learned how he applied his tinkering mind to the job of designing the oven. Andrew is not as talkative as Blair, but he's just as happy to talk. He is tall with deep-set eyes, and he wears a black brimless cap over his shortish curly salt-and-pepper hair. The few times we've met, he's had a five o'clock shadow, but there is nothing gruff about him.

"I used to spend my time pondering brick ovens," he said. Not many years into running the bakery, the original oven needed repair. This was an Alan Scott oven, the kind that launched a whole generation of microbakeries.

Alan Scott sold his plans for a reasonable price, and the materials were not expensive. People could launch a small baking business without the financial burden of a costly oven. In many cases, the oven builder helped finance start-ups by delaying payment for materials and his own labor.

Alan Scott designed these ovens to fit the concept of a village bakery. The hearths could bake about 250 loaves, a manageable day's work for a single baker. Once a community needed more bread, another baker and another oven would surface. However, making the figures work for this type of oven is tough, and bakers often move on to ovens with more efficiency and a larger capacity.

Elmore Mountain Bread was started fifteen years ago by Dave Deciucies, Andrew and Blair's neighbor and friend. Eric Blaisdell had taken over the business from Dave and built the Alan Scott oven in the house where the couple now work and live. After several years of heavy use, the oven was beyond repair. The deck was cracked, and so was the thick outer layer of concrete that added to the oven's thermal mass.

Determined to build a better oven, Andrew started researching. He found a brick oven email group and began asking questions. He consulted masons and looked at lots of designs through the Masonry Heater Association. Working with William Davenport of Turtlerock Masonry Heat, Andrew came up with a plan for an oven that could handle the amount of bread they wanted to bake. A steel-reinforced harness would support the thermal mass, and highly efficient insulation would minimize the bulk of that mass.

William Davenport built the oven for Andrew and Blair, and constructed others with similar features, including the one at Tabor Bread in Portland. The style is an affordable alternative to Llopis ovens, a model many American bakeries import from Spain.

Turtlerock is no longer in business, but Davenport's apprentice Jeremiah Church is carrying on the ideas and work that began at Elmore Mountain Bread. This oven echoes in others, too, like the one at Wide Awake Bakery. Blair was one of the people that Stefan Senders spent hours with on the phone, figuring out the best way to trap fire and heat for bread.

Elmore Mountain Bread's oven has attracted many admirers. While people come looking for information, the conversations are two-way streets and influence Andrew and Blair's thinking. Thoughts of milling began with visitors who mill for their bakeries. Jules Lomenda from Six Hundred Degrees Bakery in British Columbia, and Dave Bauer from Farm and Sparrow in

North Carolina came on separate trips. Prompts came from closer to home, too. Friends at Bread and Butter Farm near Burlington, Vermont, were milling their own flour. Farther south in Windsor, Green Mountain Flour was milling for its breads and selling flour in retail outlets. Their biggest cue, though, came from their son Phineas.

"One of the only things he ate every day was baguettes," Andrew said. "As I was doing the ordering, which was typically thirty bags of white flour and two bags of whole wheat, I realized that this was refined food. Organic, but refined."

They figured they could do better for their son, and their customers. However, switching to whole grains can be tricky, especially in a bakery with a loyal clientele. The breads resemble what has been made the whole time there's been an Elmore Mountain Bread. They've maintained many formulas, and Blair still scores the Country French loaves exactly the same way that the original owner did.

"We're dedicated to stone-ground but we didn't want our baguettes to change," said Blair. "We wanted to use our own flour without rebuilding our customer base."

They knew that sifting styles and baking habits, like pre-fermenting whole-grain flours, could help them manage the transition. They thought and talked about milling and sifting with each other and other bakers. The American mills they could buy were not as big as they wanted. Bigger stones keep the grains cooler during milling, preventing breakdown of oils. European options were better, but more expensive. Besides, Andrew liked the challenge of making a machine.

He cast his net for advice. Baker, miller, and Red Fife wheat revivalist Cliff Leir from Fol Epi in Vancouver, British Columbia, had built a mill. He sent pictures and recommendations. Andrew brainstormed with Fulton Forde and Bryn Rawylk, other bakers who also wanted to build their own mills. Over email, and without ever talking on the phone, the three worked out details for a very rustic, simple machine in a very twenty-first-century American Bread Movement fashion.

The plan was for a large grinding surface, cut with an extra set of furrows to shear off layers of bran more completely, facilitating easier sifting. Initially, Andrew and Blair talked to stonecutters in Vermont, but decided to work with Meadows in North Carolina and had a set of stones cut to their specification. The stones were bigger than any used in the company's standard mills, and would be set horizontally instead of vertically.

The metal work was more local. Friends at Iron Arts metal shop had made the door for the bakery oven and helped make the oven loader, so they fabricated the framework to house the millstones. The sifter they bought ready-made from Meadows, but Andrew built new screens to more closely regulate the process.

Unlike other tweaks they've made to their operations, the mill adds rather than subtracts labor. They're not saving any money, either. The grain cost about as much as the organic flour they were using, but the difference in product is worth it.

The flour is fresher than any they could buy, and they can control its texture and bran content, two factors that influence how flour works in dough. Commercially produced sifted stone-ground whole-grain flours run between 75 and 85 percent extraction, and Andrew is pushing the percentage of the kernel they use up to 90 percent. They can also regulate the size of the bran flakes, and choose how and when to incorporate what they've sifted off back into flour and dough. Working with pre-milled flours from a bag, their path into whole grains would be very limited.

The bakers still use white roller-milled flour to dust the bench and make focaccia; some of their restaurant clients aren't as flexible in product or price.

Andrew and Blair switched all of their loaves to home-ground flour, which meant the composition went from mostly white flour to mostly whole-grain. Because of the adjustments they made through sifting and incorporating pre-ferments, the character of the breads stayed the same. However, the flavor of the loaves increased dramatically. While many flavors in bread come from fermentation, the germ and the bran that remain in the flour also contribute to the tastes of the loaves.

The new breads have been absorbed without much of a blip. One customer complained that the bread was too "whole wheaty." Once Blair explained what they had done and why, the man was no longer mad.

They raised their prices slightly to account for the extra labor and investment. One market admired the change so much that they cut down their markup to keep the price close to similar offerings.

Twelve hundred loaves a week, and more in summer, are selling in and around Montpelier and Stowe. Maybe it's good that Blair and Andrew couldn't explain to their customers about switching flours. If they had, perhaps the new breads would have been received with suspicion. Whole grains, however morally appealing, are an obligation. Regardless of what we think we should eat, most of us head for white loaves. Blair and Andrew have managed quite a magic trick.

The magic is seeping out into the bread world. As word of what they'd done spread through the artisan baking community, people started calling about the mill, curious and inspired. Some were looking for help to build their own mills, too. Andrew applied his engineering mind and drew up formal plans. Now, he's consulting with other bakers around the country, like Josey Baker in San Francisco.

However deft Andrew and Blair are with flour, though, they can't magically summon quantities of local wheat. Much of the Vermont wheat production is already tied up. Tom Kenyon of Nitty Gritty Grain Company sells most of what he grows to Champlain Valley Milling. Ben Gleason mills his wheat himself, and Red Hen Baking Company buys a good portion of his production. So most of Elmore Mountain Bread's grains come through the same distributor they've used for some time, Hillcrest Foods. The mills they've previously used for flour, Heartland Mills in Kansas and Champlain Valley Milling in New York, now sell them whole grains. These organic wheats are grown in the wheat belts of the country.

When Andrew and Blair can get locally grown grains, the bakers are thrilled. Corn comes from Butterworks Farm, and goes into Anadama bread. The Vermont wheat they found at Rogers Farmstead gets a loaf of its own, named for the variety, Redeemer. Being able to close the loop from field to loaf was a matter of pride, especially when Blair and Andrew brought Phineas to the farm. Seeing next year's Redeemer tufty and green in the field, watching their son admire the combine, being at a farm run by another young family: These were bonds they didn't imagine when they started baking bread.

Curiously, the concept of local flour resonated more easily with consumers than fresh. The response from customers about loaves made with Vermont grains has been ecstatic; *local* is an adjective more readily understood than *fresh-milled*. The bakers would like to buy a lot of this wheat, but Nathan and Jessie Rogers are milling and selling it themselves as part of a diversified farm operation. Balancing the economics of any small business is delicate, and the finances of growing staple crops needs as much or more innovation than the tools of an artisan bakery.

If Elmore Mountain represents the next generation of bakers, these are the next generation of farmers. Rogers Farmstead is new. The couple met at work and began looking for a farm. Nathan grew up farming, and Jessie had the yen. Both of them worked in technology, and their hunt took some time. They only settled on what type of farm they would run once they had found a location and saw how their land could work. They have 135 acres, 95 of which they are farming; 75 of those are in grains. They'd like more land, and are using a combination of grains, dairy, and livestock to balance the farm's economy and ecology.

With a grant from the state's Working Land Initiative, Nathan built a grain cleaning area and storage bins in the barn. They've grown and sold wheat and oats, milling and flaking them in small tabletop machines. They bring these to farmers markets and sell grains, milk, and meat on the farm. Soon they'll set up a small Osttiroler mill. Jessie telecommutes to stabilize the family income, but every step they take is toward building a sturdy, profitable farm. This is why they have to limit how much grain they sell to anyone else.

Other on-farm mills in Vermont, begun decades ago at Gleason's Grains and Butterworks Farm, and more recently at Boundbrook Farm, were an inspiration to Jessie and Nathan. I hope their undertaking will inspire other farms. There is a need for grains in the state and in the Northeast, as Green Mountain Flour is finding.

Zach Stremlau and Daniella Malin run a mill and bakery, and they're committed to using regionally grown grains. Corn is plentiful, but sourcing other grains, especially wheat, has been a struggle.

Zach had been a baker for decades, but the inspiration for Green Mountain Flour came during a stint as a cheesemaker. Before dawn he would go out into the foggy pastures to bring the cows in for milking. A few hours later, he made cheese from that same milk. Feeling the direct connection from pasture through cows to cheese inspired Zach to bring that same direct connection to bread. A few years later, they started the bakery and mill, establishing a livelihood tied to New England's working landscape.

Green Mountain Flour sits on top of a hill at the edge of the small town of Windsor, nestled near fields and houses. A mobile oven is parked next to their home. Two other wood-fired ovens, both made from straw and clay, pop out the back and side of a converted garage. A wall divides the mixing and baking area from the mill and the sifter. Zach uses a 30-inch stone mill to make flours from wheat, spelt, rye, oats, buckwheat, and corn. Opposite the Meadows mill lie bags of grain, ready for milling, and bags of bran, ready to go to farms. Down the hill, in town, they rent storage space for more grains.

The enterprise is not the family's only source of income. Daniella works at an organization that promotes sustainability in the food system, the Sustainable Food Laboratory. She does most of the sourcing for milling grains, a search that demands both time and patience.

Supporting the region's iconic working landscape and supporting farmers to build soil health are their top priorities. Use of herbicides is out, and they will source certified organic grains if at all possible. But locally grown grains are their highest priority, so if it is a choice between non-certified but responsibly grown grain from nearby versus a certified organic crop from farther away, they will buy the local crop. The idea is to support farming and create relationships with growers near them. Setting a strict policy of organic-only would narrow their choices and limit the dialogues they could have, along with their potential to influence what is grown nearby.

Realistically, in some years, the choices for food-grade grains, regardless of the way they are grown, can be slim. A couple of bad years for wheat growing in the Northeast have made the search very tough. Daniella and Zach talk to farmers, mills, grain brokers, and extension agents, tracking down responsibly grown local crops.

One challenge is the ingrained perception of flour as an anonymous ingredient. Vermont is a leader in local food production at the grassroots and government levels, but even in a state that's wildly prejudiced in favor of local, artisanal food, raising the social profile of staple foods is tough. Outreach and education is an important part of the work that Zach and Daniella do, letting people know what is different and good about stone-ground flour and grains from nearby. People are responsive: Their pancake mix is used in schools and at inns, and their flour and bread sells at co-ops and natural foods stores.

Unfortunately, sourcing directly from farmers has its risks. One large shipment of wheat came with good test results, but made bread that didn't taste good. Samples of the grain had tested fine in the lab, but somewhere between testing and shipment the grain lost its baking qualities. The lesson?

Always taste the grain first, even if the numbers tell a good story. The couple lost a lot of money, and some ground, but they are faithful to their vision for a regional food system. Their business, and others like it, they know, will help create the grain supply they want to see.

The mill is a bridge to the land. The lack of traffic on this particular bridge shows the grain deficit in the Northeast. We've lost farmer know-how and the materials required, from seeds through storage, for growing food-grade grains. When I began looking at grain productions, I underestimated the size of this loss. I was naive in my love for flour and thought that markets created by enchanted eaters would flip a switch. Slowly, I am learning that change in farming is glacial. The pace may change, however, as a project of regional grains is receiving serious attention and resources just south of where I live.

The Hudson Valley Farm Hub is a new nonprofit center for resilient agriculture covering 1,255 acres south of Kingston, New York. The goals of the farm are broad. The target is to establish a viable food system that is economically and environmentally sound; grains are a part of this target. Research and farmer training are the main channels of the work. Through demonstration farming and education platforms, the Farm Hub is creating support for mid-sized farms, first in the Hudson Valley then, eventually, at the national level.

The venture is just beginning, but its concept has been brewing for a while. When the New World Foundation's Local Economies Project made a study of regional food production, it found a significant acreage in crops that mostly support dairies, like corn, hay, and forage. Identifying ways to foster strong farm systems, the foundation realized that the demand for grains was an obvious market to target, as was the need for training the next generation of farmers. The project drew in collaborators. The New World Foundation purchased the Gill Farm in 2013, a family sweet corn and vegetable operation, as a site for the Farm Hub, and the NoVo Foundation is providing initial funds to run the center. Farmer John Gill stayed with his land as the farm operations manager and resident farmer mentor to guide the project. Within the first growing season the Farm Hub began experimenting with various arrangements of cover crops and established its first formal research projects in grains with Cornell.

The Farm Hub is concentrating on solving problems for farmers by examining seed varieties, cultural practices, and equipment. Cornell University and the Cornell Cooperative Extension of Ulster County are research partners in a five-year grain project. Each season will test twenty to thirty varieties of barley, hybrid ryes, and winter and spring wheats in a mixture of organic and conventional plots.

Varieties that succeed will be planted to larger acreages, and as the crops grow to market scale the Farm Hub will be working with processors to test the results. The structured investigation is a tool to mitigate risks for farmers who want to break into grains but can't afford to set aside land and resources for unknown crops.

Other experiments look at intercropping grains with layers of mixed cover crops. The first year, a popular heritage corn variety named Bloody Butcher was planted in a field crop/value-added grains rotation. This kind of work looks at building soil health while supporting a farm's economy, which is always a challenge. All of the Farm Hub's work is geared toward helping farms thrive, especially midsized farms that can produce food in volume. The overarching concept is to help farms meet a triple bottom line of social, economic, and environmental viability.

Migliorelli Farms is a good example of the type of farm that needs this help. Ken Migliorelli is already pursuing these goals, managing 1,000 acres in the Hudson Valley through diversified operations. The farm began with a single acre of vegetables in the 1930s.

"My grandfather grew and peddled vegetables in the Bronx," Ken said. He still grows broccoli raab from the seeds his grandfather brought from Italy, and has expanded into a range of crops and options, including orchards, and 400 acres in vegetables alone. Ken also keeps twenty head of beef, which he sells at his own farm stands. About 80 acres are planted to hay. For the past five years, he's put hundreds of acres in grains, trying to turn some of his rotations into marketable crops.

"I always planted a green manure, but I've been trying to make some money on the cover crop," Ken said. "I've had a hard time getting DON levels below 1 ppm."

Many years, because of vomitoxin, the grains achieve only feed grade, not food grade. Wheat has been tricky, too, trying to get the right protein levels. Coppersea Distilling is buying his rye, and the barley he grew last season almost hit the mark. The crop tested below 1 ppm for vomitoxin, but the pre-harvest grain sprouting level was high, making the germination rate uneven. The crop wasn't right for maltsters, who need high levels of ungerminated grain for malting.

Altogether, Ken grows 150 types of crops and markets through wholesaling, farmers markets, and value-added foods. At one point he sold at forty different farmers markets, but he's cut back to focus on growing the wholesale side of the business. The maze of vegetable production commands a lot of his time, but he got into grains because he already had some equipment for cover crops. For a couple of weeks each summer, adding the grain harvest to his list of work makes life extra-crazy. Yet he keeps at it because he knows the value of pursuing multiple revenue streams, and he'd like to tap into the demand for local grains. The crops are also attractive because they are less labor-intensive than vegetables and fruit.

Ken's eagerness to make grains work is evident in the fact that he hosted the first on-farm variety trials of grains run by Cornell in the Hudson Valley in recent memory; possibly these were the first Cornell grain trials in the area, period. The heritage grain trials of 2014 were partially sponsored through Glenn Roberts, who has created a model of recovering seed varieties in South Carolina through farming heritage crops and marketing them through Anson Mills.

Ken is glad to see grain research at the Farm Hub, and elsewhere in the Hudson Valley. Other area philanthropists have growing interests in food security, and in supporting local farm systems and experimentation that includes grains. Ken hopes these projects will figure out the best varieties and practices for growers, and maybe get a grain elevator in place. Growers don't have the time or money to make mistakes.

Radical change is risky in any business, but especially in agriculture, where farmers are always gambling against the variables of working within a natural system. Plus, equipment is expensive. If farmers actually witness a large farm survive and profit on something other than soy and corn, it will be a powerful driver in considering changing what they plant. Centralized seed cleaning and storage facilities could also help those farmers transition, and would build a critical mass of production for specialty grain crops.

I want to believe that these investigations, and all the other farm and food supports I've seen, will work magic and dramatically change the way that most of our grain is grown and handled. But making grains work outside specialized cash-crop farming is tricky. Not just because we've lost farmer know-how, seed varieties, and infrastructure, but also because this is a battle against one of the big-daddy commodities. Unless I can summon a grains-obsessed billionaire to bankroll fleets of combines and rows of grain bins, and install mills on farms and in cities, change will come as it already has come—not wholesale, but in pieces.

People will keep studying one another and drawing on their ingenuity to build sustainable farms and food systems. Alan Scott jump-started a new old-fashioned approach to bread with his oven plans, offering an alternative route to a food that had been industrialized. Other innovators are fiddling with ovens and mills, turning dairy tanks and silo bottoms into malt systems, scaling down equipment, and deindustrializing processing. They are making tools to fit a future they are shaping.

While my tendency is to think that grains alone will save the day, grains are just a part of a complex farming picture and the changes happening as food production is relocalized. Nathan and Jessie Rogers wanted to be farmers, not grain farmers. Grains are a part of a diverse operation for them, one that includes animals and crop rotations. Farm health is also at the root of the many projects undertaken by the Dewavrin family over the last two decades. Grains are not an end for them, but part of their general pursuits. They are, in Loic's words, "trying to find the best combination of operations in the field to respect the environment as much as possible."

Rogers Farmstead and Les Fermes Longpres are building their farms by collage, piecing together equipment, some of it used or antique, with handmade solutions, as most farmers do. Nathan Rogers built storage bins from plywood and found a used Osttiroler mill at a fraction of the cost of a new one. The Dewavrins scouted abandoned roller mills across the Atlantic to suit their needs and restored them, pairing the old machines with new sifters. Both of these farms are enabling bakers in Vermont to know who grows their wheat, and how that wheat is grown, with the same care they put into their loaves.

Regional grain projects don't operate in isolation. They fit into the larger picture of sustainable agriculture. As farmers pursue more holistic methods, they are balancing ecological concerns with market realities. Farming is a gamble against seasons and soils and machines, so changing farming means changing the shape of the markets and nudging along production through purchases. Not my kind of enthusiastic retail purchases, but quantity purchasing by processors like Valley Malt in Massachusetts, Grist & Toll in Los Angeles, and the Somerset Grist Mill in Skowhegan, Maine.

Public works initiatives that mandate usage of regional grains also encourage production, prodding supply by creating demand. Greenmarket's rule for bakers developed mills and got more grains in the ground, and New York State's Farm to Brewery licensing is creating a market for malt. Entrepreneurs are starting small malthouses, and farmers and researchers are trying to find malting barley varieties suited to the region.

In Oregon, Hummingbird Wholesale fosters change by helping fund the kind of farming it wants to see and the kind of food it wants to sell. The distributor functions like the community and economic development partners involved in The Kneading Conference in Maine, and in the Skagit Valley in Washington. Washington State University's Steve Jones has built The Bread Lab to support grain growing outside the wheat belt. Steve's vision is so strong that reach of this work stretches across the country, drawing attention to the model, and support from national companies like King Arthur Flour and Chipotle.

Mainstream grain customers are exploring non-commodity options. Bigger small breweries like Sierra Nevada, Rogue, and Peak Organic all find ways to use regionally grown and malted barley. Whole Foods sells local flour from Maine Grains, and uses the mill's flour to bake local loaves distributed in and around Boston. Grist & Toll's flour is used in some Whole Foods bakeries in California. King Arthur Flour has even launched a regional product: West Coast Artisan Flour. While the interest of such companies can leverage more grain production, does it threaten the success of grassroots, community- and farm-scale projects? And is the interest more than locavore posturing?

Since King Arthur is a small business, I've had many opportunities to pose my doubts and get some good answers. Not answers spoon-fed by corporate positioners, but from conversations with individuals: Baker Jeffrey Hamelman told me that King Arthur's employee-ownership structure gave him entitlement to follow his instincts and join the Northern Grain Growers Association, where he could add a baker's insight to the work of the group.

Instructor Amber Eisler helped me see more clearly the commitment to educating home bakers and professionals. Meeting bakers' interests through education was part of the flour company's job, whether people were curious about gluten-free baking, croissants, or using local flour. Building an education center at King Arthur's headquarters in Norwich, Vermont, was part of that focus. Donating flour and instructors, first to The Kneading Conference and now to The Bread Lab, is simply an extension of it.

This is an old company with roots in Boston and a legacy of quality. King Arthur began importing European flour in 1790 to meet bakers' needs for first-rate ingredients. Almost two hundred years later, when I started buying my own flour, that reputation for quality led me to this brand. I only quit using it as I discovered local options. While the company is not trying to personally lure me back, King Arthur is trying to meet customer desires for grain of known origin without compromising on its goal of steadily delivering quality.

The company's first regionally identified flour was developed for commercial bakers who wanted a product with a geographic link to their bakery. Initially, all the wheat came from California, but the sourcing territory widened because drought limited the crop. King Arthur is investigating another line in Kansas, where it could release a family flour product and identify farms and farmers on the label. (*Family flour* is an industry term, delineating home bakers and the products geared to them.) There is interest in regional production in New York, too, as I learned from Brad Heald, King Arthur's director of mill relations.

"I would love to hitch our wagon to a regional variety that tastes good and is consistent year on year," said Brad. Most of his job is less dreamy and more realistic, focused on details of tracking actual grain supplies rather than possible stocks. Working with mills to meet projected usage, he initiates flour specifications and contracts.

Sourcing within a region is not just farm-to-table allure, but a search with simple motives. King Arthur sticks to American grain for its flours. When a continued drought in the southern Plains in 2014 limited the supply of organic bread wheat, King Arthur told its commercial bakery customers that there would be no more organic bread flour until the next American harvest. Other flour companies turned to Argentinian-grown organic wheats in the early spring.

Expenses are coming into play with grain sourcing, too. As trucking materials across the nation has become more costly, the quest for regional grains is grounded in simple economics. The Northeast may not meet the company's needs, but any percentage will be helpful.

Toward that end, Brad has been following research at Cornell and soft wheat production in New York State. While pastry wheat remains 99 percent of what is grown, he's seeing incremental changes in New York State. The need for bread wheat in the region has hard wheat varieties on farmers' radar, and researchers', too. Just above New York in Ontario, the Canadian government began funding hard wheat research at the University of Guelph. Previously, research had focused on soft wheats, so the addition is rather telling.

Of all these practical steps toward regional grain production, the most hopeful I see is the reappearance of small mills and malthouses. The machines may function out of public sight, but local flour and craft malt can introduce people to processes that have largely disappeared. That's how it worked for me. The first bite of that cookie from Wild Hive broadcast the fact that something was different.

Mills and malthouses are platforms that let bakers and brewers announce new ingredients. Elmore Mountain Bread didn't scream and shout, *Hey, we're milling!* They are rock-star bakers, but they are not a band, standing in front of a crowd with a microphone, saying, "Check out our new flour; this stuff is the bomb!" In my kitchen or teaching cooking classes, I get an audience rather often, but the pancakes I make are more powerful allies for local grains than I am.

When I go on about fresh flour, people can tell that I'm engaged in an idea, but the pancakes help convey what I'm saying. One second I look like a nut wielding a spatula. The next, my words mean something because people are eating pancakes made from organic stone-ground whole-grain flour. I still seem like a nut, but my topic makes sense.

In writing, there is a commandment: Show, don't tell. Fresh flour shows what I have told in this book. I've described the magicians who are pulling rabbits from thin air. Farmers, millers, bakers, brewers, and maltsters are drawing attention to ingredients that have become anonymous. Flour and malt did a vanishing act. I've peeled back the velvet curtain, told you what it takes to make these things. Showed you the grains all-stars who are doing some heavy lifting. You have met them, and now you need to meet their work. Find some bread and beer made with off-grid grains. Make some pancakes at your very own griddle. Your tongue can show you why all of this matters. I swear.

THE BEST PANCAKES

1 cup white whole wheat pastry flour
1 teaspoon double-acting baking powder
¼ teaspoon baking soda
½ teaspoon salt
2 eggs
½ –¾ cup milk, depending how thick you like your pancakes
1 tablespoon yogurt

Whisk together the dry ingredients, and add the liquids to the same bowl. Combine thoroughly and let rest for 10 minutes. This allows the flour to absorb liquids.

Heat a griddle until water dances on the surface, or melted butter just starts to darken. Sizzle plenty of butter, probably a teaspoon or two, on a 12-inch griddle. If this griddle is aluminum, you will be best equipped. If your aluminum griddle has a temperature dial in the handle reading COLD, READY, HOT, you will know exactly what you need to know about your heat.

Once the griddle is ready, spoon small rounds of batter on the buttered griddle. When bubbles just start to form, flip the pancake and cook it briefly on the other side. Serve with butter and yogurt, and, if you like, maple syrup.

ACKNOWLEDGMENTS

*T*his book would not have been possible without the people who have shown me their work, answered my questions, and put up with my single-minded interest in grains. I am grateful to all those who let me tell their stories.

Special thanks to the grains teams at Cornell, the Universities of Vermont and Maine, The Bread Lab, OGRIN, NOFA-NY, and the Greenmarket Regional Grains Project. Your work pushing this movement forward is inspirational.

I am also grateful to a lot of people who work behind the scenes, facilitating the work I've illustrated: Wendy Hebb, Michelle Russo, and Glenda Neff are just a few. Others contributed background information, like Richard Miscovich, Barak Olins, Ken Albala, Vern Grubinger, Andrew Ross, Paula Marcoux, and Sharon Burns-Leader. William Rubel's *Bread: A Global History* and Andrew Smith's *Eating History* were very helpful. My neighbor Howard Stoner has been a great asset. Thanks for keeping an eye on my backyard plots and toting your bicycle-powered grain mill all around town.

Many outlets let me tell the stories that led to this book. The first was From Scratch Club, a community of blogging cooks. Thank you, Christina Davis, for encouraging me to put my explorations in words. I learned a lot about grains by writing for *Lancaster Farming* newspaper, NOFA-NY's newsletter, *Edible Green Mountains*, *Edible Finger Lakes*, *SeedStock*, *Culinate*, *Zester Daily*, and *Civil Eats*. I'm happy my local paper, the *Albany Times-Union*, runs some of my essays, parts of which have made their way into this book.

I'm especially grateful for Oechsner Farms, Farmer Ground Flour, and Wide Awake Bakery. These guys are the poster children, or pinup men, of the movement to re-regionalize grain production, and helped me understand grains from the ground up. It was great to have miller Greg Mol on speed dial. Stefan Senders, baker and brainiac, fed me great bread and the ideas that began this book. Thor Oechsner, farmer and book coach, fed me facts and, more important, faith that I could tell the tales of this moment in grains. Rachel Lodder, thank you for your incredible photographs of the farm in all seasons; they helped me tremendously.

Andrea Stanley, thanks for steering my pancake mania into the path of my editor at Chelsea Green, Ben Watson.

The biggest help of all was having a family who let me follow flour back to the field. Felix, Francis, and Jack have endured my enthusiasms for grains with humor and interest. Thanks for never getting tired of eating pancakes. Felix, you are my grains pal. I love your dreams about tractors and mills. Francis and Jack, I'm glad you love plants. Thanks for teaching me about them, and for growing so much of our food. Jack, you've grown this writer. Thanks for always letting me put writing first, and sharing me with my obsessions. I couldn't ask for a more generous and understanding mate.

GLOSSARY

ANCIENT GRAINS: This is more of a marketing concept than a classification. The name suggests grains that date from very early in the domestication of crops. Ancient types of wheat include einkorn, emmer, and Khorasan.

CLASSICAL PLANT BREEDING: The deliberate crossing of plants from chosen plant parents, followed by selection of new varieties with desirable properties from the resulting populations.

DON: Deoxynivalenol, a vomitoxin that may result from contamination with the *Fusarium* fungus. Wheat for human consumption is not permitted to have DON levels above 1 ppm because of the potential to cause human sickness (vomiting).

GLUTEN: A diverse group of two storage proteins in the wheat grain: gliadin and glutenin. These two storage proteins make up gluten as dough is hydrated, and provide the elasticity and extensibility needed for making yeasted breads. Carbon dioxide is trapped by this network of linked glutenins and gliadins and causes the bread to rise. These storage proteins are present in varying degrees in all wheats, including einkorn, emmer, and spelt, as well as to a lesser extent in barley and rye.

GMOs: Genetically modified organisms are made through the insertion of genetic material using techniques of genetic engineering, rather than manual crossing of two plants. In grains and legumes, GMO corn and soy are common. GMO wheat is in the test phase, but not yet on the market.

GRAINS: The edible seeds of certain plants from the grass family.

HARD WHEATS: These are generally used for bread, because of their higher protein levels. The hardness refers to the quality of the endosperm.

HERITAGE OR HEIRLOOM GRAINS: Crops that are products of human selection and that were developed before professional breeding programs existed.

MODERN GRAINS: Varieties developed after 1950 through breeding programs. These plants generally have shorter straws (stalks) because of the incorporation of dwarfing genes.

PSEUDO-CEREALS: Plant foods that are used for their flour but that are not true cereals. Buckwheat, amaranth, and other foods that fit into the eating category of grains but do not share any plant relations to true grains.

SOFT WHEATS: Soft wheats are mostly used for pastry, crackers, and other non-bread foods.

INDEX

INDEX

ABOUT THE AUTHOR

*A*my Halloran has been following the revival of regional grain production in the Northeast for several years. She writes about food and agriculture for farming newspapers, cooking websites, and regional magazines. Her involvement in local food systems began with the Troy Waterfront Farmers Market in upstate New York, which bloomed under her care to a fifty-vendor year-round marketplace with more than a thousand weekly shoppers. She works with friends and neighbors to change the foodscape in her city, teaching classes in cooking, baking, and food justice, and volunteering at a youth-powered farm. She likes to cook for a lot of people, whether at community meals or managing a soup kitchen that incorporates as much fresh food into the menu as possible. She never tires of pancakes.

the politics and practice of sustainable living

CHELSEA GREEN PUBLISHING

Chelsea Green Publishing sees books as tools for effecting cultural change and seeks
to empower citizens to participate in reclaiming our global commons and become
its impassioned stewards. If you enjoyed reading *The New Bread Basket*, please
consider these other great books related to food and health.

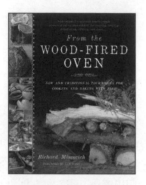

FROM THE WOOD-FIRED OVEN
*New and Traditional Techniques for
Cooking and Baking with Fire*
RICHARD MISCOVICH
9781603583282
Hardcover • $44.95

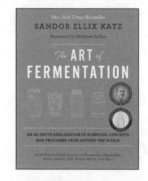

THE ART OF FERMENTATION
*An In-Depth Exploration of Essential
Concepts and Processes from around the World*
SANDOR ELLIX KATZ
9781603582865
Hardcover • $39.95

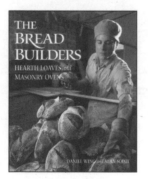

THE BREAD BUILDERS
Hearth Loaves and Masonry Ovens
DAN WING and ALAN SCOTT
9781890132057
Paperback • $35.00

HOME BAKED
*Nordic Recipes and Techniques
for Organic Bread and Pastry*
HANNE RISGAARD
9781603584302
Hardcover • $39.95

CHELSEA GREEN PUBLISHING
the politics and practice of sustainable living

For more information or to request a catalog,
visit **www.chelseagreen.com** or
call toll-free **(800) 639-4099**.